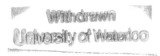

Light Scattering in Liquids and Macromolecular Solutions

Light Scattering in Liquids and Macromolecular Solutions

EDITED BY
V. DEGIORGIO

Gruppo Nazionale Elettronica Quantistica e Plasmi del CNR
Milan, Italy

M. CORTI AND M. GIGLIO

Centro Informazioni Studi Esperienze
Segrate, Milan, Italy

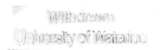
PLENUM PRESS   •   **NEW YORK AND LONDON**

Library of Congress Cataloging in Publication Data

Workshop on Quasielastic Light Scattering Studies of Fluids and Macromolecular Solutions, Milan, 1979.
    Light scattering in liquids and macromolecular solutions.

    Proceedings of the workshop held May 11—13, 1979, at the Centro informazioni studi esperienze (CISE)
    Includes index.
    1. Liquids—Optical properties—Congresses. 2. Macromolecules—Optical properties—Congresses. 3. Quasielastic light scattering—Congresses. I. Degiorgio, V. II. Corti, M. III. Giglio, M. IV. Centro informazioni studi esperienze, Segrate, Milan. V. Title.
QC145.4.O6W67 1979   530.4'2   80-20472
ISBN 0-306-40558-X

Proceedings of the Workshop on Quasielastic Light Scattering Studies of Fluids and Macromolecular Solutions, held at the Centro Informazioni Studi Esperienze (CISE), Segrate, Milan, Italy, May 11—13, 1979.

© 1980 Plenum Press, New York
A Division of Plenum Publishing Corporation
227 West 17th Street, New York, N.Y. 10011

Printed in the United States of America

PREFACE

    This volume contains most of the papers presented at
the "Workshop on Quasielastic Light Scattering Studies of
Fluids and Macromolecular Solutions" held at CISE, Segrate
(Milano), Italy, from 11 to 13 May, 1979. Quasielastic light
scattering (also called self-beating spectroscopy or intensity
correlation spectroscopy) is the technique, introduced by
Benedek and coworkers and by Cummins and coworkers about 15 years
ago, by which dynamical information about a scattering medium is
obtained through the measurement of the power spectrum (or the
intensity correlation function) of the laser light scattered from
the medium. The technique received in the early seventies a
considerable impulse from the development of real-time fast
digital correlators.

    The aim of the Workshop was to bring together a selected
number of researchers in order to discuss recent developments in
quasielastic light scattering and related optical methods and to
report about new applications of the technique in physics,
chemistry and biology.

    The first two days of the meeting were devoted to the oral
presentations of papers. In the third day an informal session
was held which included a very lively and thorough discussion of
the highlights of the Workshop. Most part of the technical
problems were debated during the informal session and during the
subsequent visits to the light scattering laboratory in CISE.

    All the contributions are grouped with some degree of arbitrari-
ness into the four topics which formed the underlying structure
of the program. The first topic is the study of interactions among
colloidal particles in solution. Light scattering appears to be
a very powerful method for this kind of study which is rich in
fundamental and practical implications. We have also included in
this first group of papers an interesting investigation about
the internal degrees of freedom of polymers in solution. The
second topic is the behaviour of amphiphilic molecules (such
as, for instance, detergent molecules or phospholipids) in

aqueous solutions. These molecules may associate spontaneously
to form various structures, micelles, vesicles, microemulsions,
bilayers, which are very conveniently studied by quasielastic
light scattering techniques. It should be mentioned that, besides
those reported in this book, important light-scattering studies
of micellar solutions were presented at the meeting by G.B.
Benedek, J. Kratohvil, A. Lips, and G. Porte. The second group of
papers includes also a very sophisticated light scattering
experiment about conformational changes of hemoglobin, and a study
of the motility of small microorganisms. The third topic includes
phase transitions in thermodynamic systems and convective instabilities
in single-and two-component liquids. It is a remarkable fact
that the same optical techniques may be used to study both kind
of systems, and that the experimental results show many
conceptual similarities. The fourth topic is not homogenous
with the previous ones because it is a technique, namely
forced Rayleigh scattering, discovered several years ago
and now applied in many dynamical studies of liquids and
macromolecular solutions. Such a technique can be considered
complementary to quasielastic light scattering.

The Workshop was jointly sponsored by the Italian National
Council of Research (CNR), through the Gruppo Nazionle .
Elettronica Quantistica e Plasmi (GNEQP), and by the Centro
Informazioni Studi Esperienze (CISE). We wish to express our
appreciation to the Scientific Council of the GNEQP-CNR for
the financial support and to the CISE staff for making available
free of charge the facilities of CISE and for their cooperation
during the preparation of the meeting. We thank in particular
Mrs. D. Fano for her precious assistance through the organization
of the meeting and during the Workshop days.

V. Degiorgio

M. Corti

M. Giglio

CONTENTS

III: PHASE TRANSITIONS AND HYDRODYNAMIC INSTABILITIES

IV: FORCED RAYLEIGH SCATTERING

CONTENTS

SOME EXPERIMENTS USING QUASIELASTIC LIGHT SCATTERING

P. N. Pusey

Royal Signals and Radar Establishment
Malvern
Worcestershire, U.K.

INTRODUCTION

This paper reviews three loosely related sets of experiments
in which quasielastic light scattering (QELS) was used to study
liquid suspensions of particles. The aim here is to give an
overview of the work with emphasis on the physical principles
underlying the experiments as well as their implications for future
research. The reader is referred to the original published
reports for quantitative details.

First (section 2) we review several years of research on
suspensions of charged, spherical, colloidal particles. This
work is starting to provide a fairly complete picture (at least
qualitatively) of the effects of strong interparticle interactions
on the dynamics of dilute particle suspensions. One of the last
pieces of the puzzle has recently fallen into place with the
recognition by Weissman (1) (reviewed in section 2.2) that
polydispersity (that is a distribution of particle sizes) can have
a much greater effect on light scattering measurements on interacting systems than was previously realized. Although this matter
was not discussed at the Milan meeting, it resolves an apparent
disagreement between experiment and theory (which did figure in
the meeting) and is therefore relevant to those proceedings.
Suspensions of charged particles are also discussed elsewhere in
this volume from the viewpoints of experiment (2), theory (3), and
computer simulation (4).

The studies referred to above used a scattering volume V big
enough to contain a large number <N> of particles whose spatial
correlations extended over distances much less than $V^{1/3}$. In this

1

limit the scattered electric field is a complex Gaussian random
variable and the quantity measured by self-beat QELS, its fourth-
order correlation function (the intensity correlation function),
can be written in terms of its second-order function (equation 1)
(5). Here the intensity fluctuations in the scattered light
result from interference between the elementary fields scattered
by different particles and the measurements can be interpreted
entirely in terms of the second-order particle properties. However
it has long been recognized that, by using small scattering
volumes (non-Gaussian scattering), it should be possible to
measure more complicated properties of the particle systems,
for example, time-dependent fourth-order spatial correlation
functions (section 3.3). Thus experiments in the non-Gaussian regime
should provide more information than is obtained from Gaussian
scattering (5). The first experiment of this type was performed
by Schaefer and Berne (6) in 1972 who studied the light scattered
by a small volume V of a suspension of non-interacting spheres.
They found an extra, non-Gaussian contribution to the intensity
correlation function, of relative magnitude $<N>^{-1}$ (which therefore
becomes negligible when $<N>$ is large), which could be interpreted
in terms of fluctuations in the instantaneous number of particles
in V. Such number fluctuation experiments have since been
exploited to study the dynamics of motile micro-organisms (7) and
to determine the molecular weight of a nucleic acid (8)*.

The experiments to be discussed in sections 3 and 4 are
extensions of the Schaefer-Berne experiment, motivated by a
desire to investigate further the potential of scattering experiments
in the non-Gaussian regime. First, in section 3, we consider non-
Gaussian scattering by interacting, spherical particles (the same
as those discussed in section 2). For the experimental configuration
used, the non-Gaussian term can again be attributed to number
fluctuations whose magnitude is, in this case, greatly reduced by
the replusive interactions.

In section 4 we turn to non-Gaussian scattering by non-interacting, non-spherical particles. Here the non-Gaussian term depends
on orientational fluctuations, as well as on number fluctuations,
through the fourth-order correlation function of the field amplitude
scattered by a single particle. This is a more sensitive function
of particle size and shape than the second-order function normally
measured in the Gaussian regime. A feature of this experiment was
the measurement of cross-correlations between the signals received
from two spatially-separated detectors. When the detectors were at

*The number-fluctuation technique of fluorescence correlation
 spectroscopy, developed independently by Webb and co-workers (35,
 36), is also finding wide application.

scattering angles + 90° and − 90°, the orientational part of the cross-correlation function had a zero-time value <u>smaller</u> than its asymptotic long-time value of one, implying an <u>anti</u>-correlation between the signals. This observation has a simple qualitative explanation given in section 4.3.

Some implications and possible further extensions of these non-Gaussian experiments are discussed in sections 3.3 and 4.4.

2.   GAUSSIAN SCATTERING BY INTERACTING SPHERICAL PARTICLES

In this section we review briefly the current situation with regard to quasielastic light scattering studies of dispersions of interacting char ed particles. These are spherical colloidal aggregates of polystyrene which,in an aqueous environment, can carry a large charge due to ionization of surface groups. At very low ionic strengths, achieved by the use of ion-exchange resins, the range of the repulsive shielded Coulombic interactions can be as large as 1μm, comparable in magnitude to both the typical interparticle spacing and the wavelength λ of visible light, and many times larger than typical particle diameters (<100nm)(9−11). With strong enough interactions the particles can actually take on a "solid-like" spatial arrangement i.e. they form a "colloidal crystal"(12−14). With weaker interactions a "liquid-like" state with significant short-range order is found and the present discussion will be limited to this case. Here structure factors, obtained from the angular dependence of the average intensity of the scattered light, show a form very similar to that obtained for simple atomic liquids by the scattering of neutrons or X-rays (9−11).

In section 2.1 we proceed on the assumption that the particles are identical. In section 2.2 the effects of particle poly-dispersity are considered. Throughout section 2 it is assumed that the number of particles in the scattering volume V is large and that $V^{1/3}$ is much greater than the spatial range of particle correlation, so that the scattered field is a complex Gaussian random variable.

2.1   Review of Selected Experimental Data (assuming identical particles)

Photon correlation quasielastic light scattering measures the correlation function of the scattered light intensity I. For Gaussian light this quantity is related to $g^{(1)}(K, \tau)$, the normalized correlation function of the scattered field amplitude E (note I = $|E|^2$)(5):

$$\langle I(K,0)\ I(K,\tau)\rangle/\langle I(K)\rangle^2 \ = \ 1 \ + \ c\left[g^{(1)}(K,\tau)\right]^2, \tag{1}$$

where c is a constant of order one. For single scattering by identical spherical particles

$$g^{(1)}(K,\tau) = F(K,\tau)/S(K) \tag{2}$$

where the intermediate scattering function or dynamic structure factor $F(K,\tau)$ is (9)

$$F(K,\tau) \equiv \frac{1}{N} \sum_{i=1}^{N} \sum_{j=1}^{N} \langle \exp i\underline{K} \left[\underline{r}_i(0) - \underline{r}_j(\tau) \right] \rangle . \tag{3a}$$

Here $\underline{K}$ (magnitude K) is the usual scattering vector, N is the (large) number of particles in V, $\underline{r}_i(t)$ is the position of particle i at time t and τ is the correlation delay time. The static structure factor $S(K)$ is

$$S(K) \equiv F(K,0). \tag{3b}$$

Figure 1 shows semi-logarithmic plots of the field correlation functions obtained from a sample of interacting spheres of radius about 25 nm at a volume fraction of about 10^{-3} (i.e. a dilute sample despite the strong interactions) (15). This sample showed marked liquid-like structure with a main peak in $S(K)$ at $K_{max} \approx 2 \times 10^5$ cm^{-1} (scattering angle 73°). The plots in figure 1 are as a function of $K^2\tau$ for three values of K :(1) $K > K_{max}$ where $S(K) \approx 1$, (2) $K \approx K_{max}$, $S(K) \approx 1.6$ and (3) $K \ll K_{max}$, $S(K) \approx 0.1$. In the absence of interactions the data would all lie on the same straight line (4 in figure 1) of slope D_o, the "free-particle" diffusion coefficient; thus the effect of the interactions is striking. The data divide naturally into two time regimes (15,16): at short times the decay is strongly K-dependent whereas at longer times it becomes roughly exponential (linear in the semi--log plots) and shows less K-dependence. The data were analyzed to obtain the initial and final slopes of the correlation functions and these are plotted in figure 2 as reciprocal effective diffusion coefficients, $\bar{D}$ and D_L respectively.

We discuss first the initial slopes $\bar{D}$ whose reciprocal is seen to follow closely the form of the static structure factor $S(K)$ (figure 2). This observation has a simple explanation which has been given by several authors (16-18) and can be discussed qualitatively in terms of the forces acting in the system and their effects on the particle velocities (6). Three types of force act on a particle, the usual Brownian force due to collisions with the solvent molecules, the direct Coulombic interaction force between particles and the indirect hydrodynamic force due to the fact that the particle moves in a velocity field in the fluid set up by the motions of the other particles. In a dilute system, such as that considered here, the hydrodynamic forces are small enough

$\ln \left[C^{\frac{1}{2}} g^{(1)}(K,\tau) \right]$

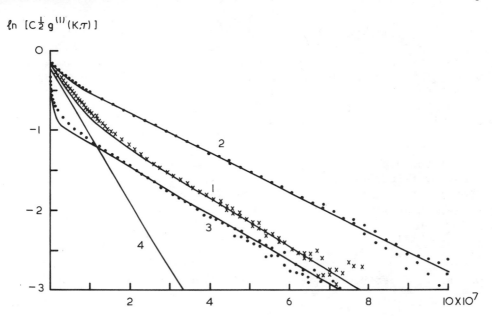

$$K^2 \tau \quad (S \ cm^{-2})$$

Fig. 1.   Correlation functions for interacting Brownian particles
for different scattering vectors relative to the position
K_{max} of the peak in $S(K)$:   1, $K > K_{max}$; 2, $K \approx K_{max}$; 3, $K \ll K_{max}$
Points are data, lines are sums of two exponentials chosen
to fit initial and final slopes.   In the absence of
interactions all data should lie on line 4. (Taken from
reference 15).

to be neglected compared to the others though this is certainly
not the case in more concentrated suspensions. The Brownian
force is strong and rapidly fluctuating and leads to large rapidly –
varying fluctuations in the particle velocity (characteristic
fluctuation time τ_B). By contrast the velocity fluctuations
associated with the interaction force are small but long-lived
(characteristic fluctuation time τ_I). On a timescale $\tau_B \ll t \ll \tau_I$,
therefore, a particle moves as an essentially free diffuser;
interactions are only felt at longer times. This picture leads
to an expression for the initial decay of $F(K,\tau)$:

$$F(K,\tau) = S(K) - D_o K^2 \tau + O(\tau^2) \tag{4}$$

where D_o is the "free-particle" diffusion coefficient given by
the usual Stokes-Einstein expression. Normalization of (4) gives

$$g^{(1)}(K,\tau) = 1 - \left[D_o/S(K)\right] K^2\tau + \dots \qquad (5)$$

so that the reciprocal effective diffusion coefficient associated with the initial decay is expected to show the same K-dependence as S(K) as is seen in the data (9,10) (figure 2). Thus the effect of the interactions on the short-time dynamics, though dramatic when plotted as in figures 1 and 2, is seen to arise from a trivial normalization.

Turning now to the more interesting longer-time behaviour, we consider first the large-K regime ($K > K_{max}$) where $S(K) \approx 1$. In this limit it can be shown that the dominant contribution to $F(K,\tau)$ is due to single-particle motions (15) i.e.

$$F(K,\tau)\big|_{K \gg K_{max}} \approx F_S(K,\tau) = <\exp i\underline{K}.\left[\underline{r}(0) - \underline{r}(\tau)\right]>, \qquad (6)$$

which, with certain reasonable assumptions, can be written

$$F_S(K,\tau) = \exp\left[-\frac{K^2}{6} <\Delta r^2(\tau)>\right]. \qquad (7)$$

Here $F_S(K,\tau)$ is the <u>self</u> intermediate scattering function and $<\Delta r^2(\tau)>$ is the mean-square displacement of a single particle in time τ. Thus a measurement of $F(K,\tau)$ for $K \gg K_{max}$ provides an estimate of $<\Delta r^2(\tau)>$. Its observed behaviour (curve 1 of figure 1) can be explained in terms of the discussion of the previous paragraph. Initially a particle sets out on a diffusive random walk in response to the Brownian forces as if it were free so that

$$<\Delta r^2(\tau)> = 6D_o\tau, \quad \tau_B \ll \tau \ll \tau_I. \qquad (8)$$

However after a while, in whatever direction it has moved, the particle starts to "feel" a weak, but only slowly varying, repulsive force due to its neighbours which tends to retard its motion so that $<\Delta r^2(\tau)>$ increases more slowly than predicted by (8). Of course, because the spatial arrangement of particles is "liquid-like", the particle will eventually find itself with a new set of neighbours. On a long enough timescale, therefore, the particle will still perform a diffusive random walk through the suspension,

$$<\Delta r^2(\tau)> = 6D_L^S\tau, \quad \tau \gg \tau_I, \qquad (9)$$

but with a macroscopic self-diffusion coefficient D_L^S (the long-time slope of curve 1, figure 1, and the $K > K_{max}$ value of D_L in figure 2) which can be several times smaller than D_o due to the retarding effect of the repulsive interparticle forces. This approach can be quantified in terms of a velocity autocorrelation function having two components due respectively to Brownian and interaction forces (15, 19) and is also discussed more formally by Hess (3).

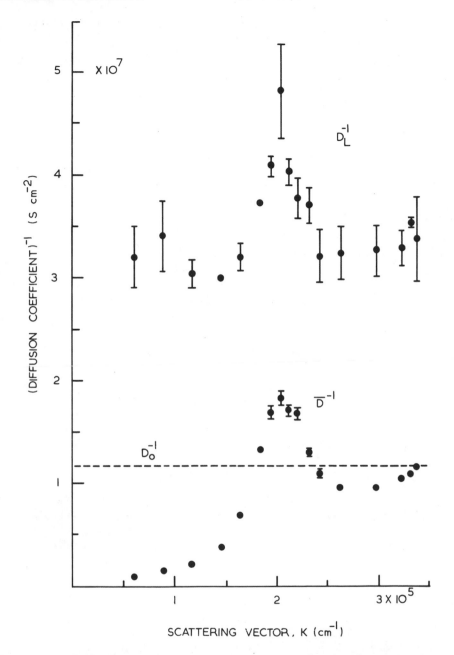

Fig. 2. Reciprocal effective diffusion coefficients obtained from
initial ($\bar{D}$) and final (D_L) slopes of curves in figure 1.
In the absence of interactions, there would be no difference
between $\bar{D}$ and D_L and all data should lie on dashed line.
(Taken from reference 15).

We now consider the long-time decay of $F(K,\tau)$ at small K where $S(K) \approx S(0) \ll 1$. Here it is convenient to recognize $F(K,\tau)$ as the autocorrelation function of

$$\sum_i \exp \left[i\ \underline{K}.\underline{r}_i(t) \right],$$

the spatial Fourier transform of the particle number density. In the small K limit, K^{-1} is many times greater than the mean inter-particle spacing so that light scattering effectively measures the decay of a macroscopic sinusoidal variation in concentration. To understand qualitatively the mechanism of this decay we consider two extreme limits. Firstly assume the particles to be non-interacting. Then the average decay of a macroscopic fluctuation will involve the actual transport of particles from the high density region to the low density region i.e. motion over a distance K^{-1} with time constant $(D_0 K^2)^{-1}$, as expected. By contrast, consider a system in the "solid-like" state. Here the particles are constrained by strong interparticle forces to the neighbourhood of lattice sites though, due to thermally-induced motions, density fluctuations still occur exactly as with atoms in a simple crystal. The decay of a long-wavelength Fourier component of such a fluctuation can then proceed by a kind of "breathing" or "overdamped phonon" mode in which lattice planes in the high-density region move slightly apart in response to the repulsive inter-particle forces whereas planes in the low-density region are pushed closer together. This mechanism involves particle motions over distances less than the interparticle spacing (and therefore much less than K^{-1}) and should lead to a decay much more rapid than that for non-interacting particles. In a liquid-like system the situation is more complicated but a rapid decay would still be expected if there is significant short-range order (see article by Hess (3) for a detailed discussion of the theories).

In terms of figure 2, this implies that, as $K \rightarrow 0$, the upper curve, D_L^{-1}, should drop down towards the lower curve D^{-1}. Our data (15) which have been confirmed by other workers (10) (and by unpublished experiments of our own (20) at very small K), show no evidence of this drop; indeed $D_L(K \rightarrow 0)$ appears to become constant at a value very similar to the large-K, self-diffusion value D_L^S (figure 2). The existence of this low-K, slow mode, not predicted by the theory mentioned above, constitutes the apparent disagreement between experiment and theory referred to in section 1. As mentioned there this difficulty has recently been resolved in terms of particle polydispersity by Weissman whose treatment is reviewed in the next section.

2.2 The Effect of Polydispersity

It is well known that, for a system of non-interacting particles, quasielastic light scattering is quite insensitive to the existence

of a small degree of particle polydispersity (21). This is because
the sum of a number of exponentials with slightly different decay
constants (each arising from a single size species of non-interacting particles) is not too different from a single exponential
with an appropriately defined mean decay constant. Probably
because of this, early work (9,15) on interacting particles tended
to assume that possible effects of polydispersity would be small.
However Weissman (1) has recently pointed out that even a small
degree of polydispersity can have profound effects on light
scattering from strongly interacting particles and is almost
certainly responsible for the slow, small-K mode observed
experimentally in the decay of F(K,τ) (and described in section
2.1). In this section we review this work and comment on its
implications.

In the presence of particle polydispersity we must retain
scattering amplitudes in equation (3) (which disappear through
normalization for identical particles) since particles of different
size have different scattering powers. Thus the measured dynamic
structure faction $F^M(K,\tau)$ is

$$F^M(K,\tau) = \frac{1}{N\langle a^2\rangle} \left\langle \sum_{i=1}^{N}\sum_{j=1}^{N} a_i a_j \, \exp i\underline{K}\cdot\left[\underline{r}_i(0)-\underline{r}_j(\tau)\right]\right\rangle \qquad (10)$$

where a_i is the scattering amplitude of particle i and the angular
brackets now include an averaging over the particle size distribution
as well as the thermal average. For simplicity, we consider only
spherical particles so that the $\{a_i\}$ are independent of time.

It was pointed out some time ago that the effect of polydispersity on the short-time decay of $F^M(K,\tau)$ can be calculated
quite simply (15). Then D_O in equation (5) is replaced by
$\langle D_O\rangle$, the usual mean diffusion coefficient weighted by the scattering
powers a_i^2 of the individual species constituting the particle size
distribution. The derivation of this result is outlined in
Appendix 1.

In the general case simplification of (10) is not possible and
the dynamics of the system must be specified in terms of a large
number of time-dependent partial pair correlation functions. However
Weissman has described an approximate but valuable simplification
which follows if specialization is made to "paucidisperse" systems,
that is systems for which the standard deviation of the particle size
is much less than the mean. We recognize that particle mobilities
go as particle radius R (Stokes-Einstein), particle charge, which
determines the interaction, goes as R^2 (for a fixed surface
charge density) whereas, at least for particle size $<< \lambda$, the
scattering amplitudes $\{a_i\}$ go as R^3. Weissman's simplification

involves assuming that the particle correlations and dynamics are
unaffected by small differences in particle size so that we need
only consider the effect of particle size on the scattering
mplitudes $\{a_i\}$. Then the averages over the $\{a_i\}$ and the phase
factor in (10) can be separated to give:

$$F^M(K,\tau) = \frac{<a>^2}{N<a^2>} \sum_i \sum_j <\exp i\underline{K}. \left[\underline{r}_i(0)-\underline{r}_j(\tau)\right]>$$

$$+ \left[\frac{<a^2> - <a>^2}{<a^2>}\right] <\exp i\underline{K}. \left[\underline{r}(0)-r(\tau)\right]>$$

$$= \frac{<a>^2}{<a^2>} F^I(K,\tau) + \left[\frac{<a^2> - <a>^2}{<a^2>}\right] F_S^I(K,\tau) \qquad (11)$$

where $F^I(K,\tau)$ and $F_S^I(K,\tau)$ are respectively the "ideal" full and
self dynamic structure factors which would obtain in the absence
of polydispersity. In terms of static structure factors, the
$\tau = 0$ limit of (11) gives

$$S^M(K) = \frac{<a>^2}{<a^2>} S^I(K) + \frac{<a^2> - <a>^2}{<a^2>} \qquad (12)$$

Several features of equations (11) and (12), which have both
advantageous and disadvantageous implications for the study of
interacting particles by quasielastic light scattering, are
immediately apparent:

(i) Equation (11) shows that a self term, determined by single-
     particle diffusion, appears in the measured dynamic structure
     factors at all values of K. This explains, at least qualit-
     atively, the observation of the small-K, slow mode seen in the
     decay of $F^M(K,\tau)$ (figures 1 and 2). This slow mode is now
     identified with $F_S^I(K,\tau)$ in (11). This identification is
     supported by the fact that the decay constant of this mode is,
     to within experimental error, the same as the slow decay
     constant for $K \gg K_{max}$ (figure 2) which has already been
     attributed largely to self diffusion since, in this limit
     $F^I(K,\tau) \to F_S^I(K,\tau)$ (section 2.1).

(ii) On the negative side, the existence of two terms in equations
     (11) and (12) means that measurements of $F^M(K,\tau)$ and $S^M(K)$
     on systems which are, in practice, always polydisperse to some
     degree cannot be directly compared with $F^I(K,\tau)$ and $S^I(K)$, the
     quantities usually obtained from theories and computer
     simulations. It will be necessary to make a "polydispersity
     correction" which will, in general, require detailed

knowledge of the particle size distribution. (However, see below, for a simplified correction in the case very narrow particle size distributions.)

(iii) On the positive side, these considerations imply that it will frequently be possible to use quasielastic light scattering to study <u>self</u> diffusion in strongly interacting systems even when the $\overline{K>K_{max}}$ range is not available experimentally. This circumstance occurs for concentrated suspensions of small particles (proteins, microemulsions etc.) where the mean interparticle spacing is much smaller than K^{-1} so that, over the whole range 0-180° of scattering angles, the structure factor S(K) hardly deviates from its asymptotic value S(0). This newly-realized potential of the technique is valuable since, although the collective behaviour described by $F^1(K,\tau)$ is of great theoretical interest, it can be argued that self diffusion is more important from a practical point of view because it determines the rate at which matter is actually transported through the system. Indeed recent measurements on water-in-oil microemulsions at high volume fractions of water have shown the existence of such a slow mode in $F^M(K,\tau)$ (22,23). If this is attributed to self-diffusion then it should be possible to deduce both the self-diffusion coefficient and some estimate of the size distribution of the water droplets. For microemulsions, these quantities, the latter in particular, are diffcult to determine by other means.

(iv) The form of equations (11) and (12) is exactly the same as occurs in inelastic neutron scattering (24). There different scattering amplitudes arise from both the distribution of isotopes constituting a given element and the interaction of the neutron with the distribution of nuclear spin states. The first term in (11) is attributed to coherent scattering and the second term to incoherent scattering. As Weissman(1) points out in the context of light scattering, the coherent scattering is due purely to fluctuations in particle number density which, as indicated in section 2.1, can, for small K, decay rapidly by relatively local motions of the particles. Incoherent scattering is due to "polydispersity fluctuations". If, for example, there is a greater than average number of large particles in some macroscopic region of the sample, this can only decay by the slow process of self diffusion, the actual transport of some of these particles out of the region. A small degree of polydispersity thus provides a natural "tag" allowing the measurement of self diffusion*.

*These considerations show the error of statements by this author in previous papers (15,19) to the effect that there is no light-scattering equivalent of incoherent neutron scattering.

For particles much smaller than λ in size, the scattering amplitude a_i is proportional to the polarizability of the particle which is in turn proportional to its volume. Thus, from (12), the ratio X of incoherent to total scattering is

$$X = \left[1 - \frac{<R^3>^2}{<R^6>} \right] S^M(K)^{-1} \quad . \tag{13}$$

For the sample which gave figures (1) and (2) the radius moments are known from electron microscopy, giving $<R^6>/<R^3>^2 \approx 1.23$ (9). In Table 1 we list $S^M(K)$, determined from the initial slope of the correlation functions in figure 1 (and corrected approximately for multiple scattering as outlined in reference 15), the ratios X calculated from equation (13) and the fractional strengths F_L of the slow mode of the correlation functions determined by analysis of the data (reference 15, Table 1). For $K << K_{max}$ we find the unphysical result X>1. This probably represents a breakdown of the assumptions on which (13) is based, namely that particle interactions and dynamics are totally unaffected by the distribution of particle sizes (1). Nevertheless the calculation shows that it is plausible that the low-K slow mode can be entirely accounted for by incoherent scattering, so that the theoretically predicted rapid decay of $F^I(K,\tau)$ is not inconsistent with the data. On the other hand for $K \gtrsim K_{max}$ the predicted strength X of the incoherent scattering is several times smaller than the experimentally measured strength F_L of the slow mode, some of which must therefore be rightly attributed to $F^I(K,\tau)$.

TABLE I

Relative strengths of the slow modes in $F^M(K,\tau)$:X, due to polydispersity (equation 13) and F_L from data analysis.

$K(cm^{-1})$	$S^M(K)$	X	F_L
0.87×10^5	0.16	1.17	0.57
2.12×10^5	1.81	0.10	0.87
3.22×10^5	0.89	0.21	0.61

To conclude this section we point out that a further simplification can be made for truly paucidisperse systems. We write
$\Delta R = R - <R>$ and expand the n'th moment of R in powers of $\Delta R/<R>$,
assumed to be small:

$$\frac{<R^n>}{<R>^n} = 1 + \frac{n(n-1)}{2}\ \frac{<\Delta R^2>}{<R>^2} + 0\left(\frac{<\Delta R^3>}{<R>^3}\right) , \qquad (14)$$

so that

$$\frac{<R^6>}{<R^3>^2} \approx 1 + 9\sigma^2$$

where $\sigma = \{<\Delta R^2>/<R>^2\}^{\frac{1}{2}}$ is the standard deviation of the particle
size relative to the mean. Then (13) becomes

$$X \approx 9\sigma^2/S^M(K) \qquad (15)$$

which enables one to obtain a rough estimate of the importance
of incoherent scattering simply from knowledge of σ and $S^M(K)$.
The commonly-used Dow polystyrene spheres of diameter 0.091µm and
0.109µm have quoted standard deviations σ of about 0.064 and 0.025
respectively (compared to $\sigma \approx 0.19$ for the sample discussed in
section 2.1(9)). Taking $S^M(K) = 0.1$ gives $X \approx 0.37$ and 0.06, so
that a small, but not entirely negligible, degree of incoherent
scattering is expected even with these samples.

3.   NON-GAUSSIAN SCATTERING BY INTERACTING SPHERICAL PARTICLES

3.1   The Experiment (25)

Two dilute aqueous suspensions of Dow polystyrene spheres
(quoted diameter 0.091µm) were prepared at similar concentrations.
Sample A was in a quartz cell containing a few beads of mixed-bed
ion exchange resin whereas sample B was in a pyrex cell without
ion exchange resin. About two weeks after preparation, measurements
were made of the angular dependence of the average scattered
intensity (figure 3). For Sample B, no structure is evident in
the plot apart from a slight fall-off with increasing scattering
angle due to intraparticle interference. We conclude that sufficient
residual electrolyte was present in this sample to shield the
particle charges so that particle interactions were negligible.
By contrast the plot for sample A shows significant liquid-like
structure due, presumably, to the development of long-ranged
forces following the removal of any residual electrolyte by the
ion exchange resin.

Photon correlation measurements were made of the intensity
correlation function of light scattered from a small volume V of

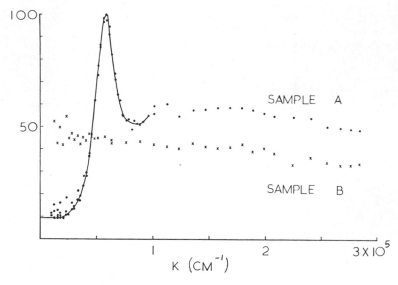

Fig. 3. Average intensity (arbitrary units) versus scattering
         vector for "interacting" sample A and "non-interacting"
         sample B. (Taken from reference 25).

these samples. The laser beam was focussed into the sample cells
by a microscope objective to provide a diffraction-limited waist
of about 2μm. A second objective,placed at scattering angle
90° to the beam, projected a magnified image of this focal region
onto a small aperture in front of the photomultiplier tube.
In this way, scattered light was collected from a volume element
in the sample of magnitude about 8μm^3 which contained about five
particles on average.

     The intensity correlation functions obtained in this way are
shown in figure 4. For both samples the functions show an
initial rapid decay, complete in 2 ms, followed by a slower
decay extending to about one second. We attribute the rapidly
decaying mode to the usual fluctuations which result from the
changing interference between the light fields scattered by
different particles and are described by the second term in (1).
Of interest to us here are the slower modes which are associated
with fluctuations in the number of particles in the scattering
volume. We see immediately that, although the samples are of
similar concentrations, the magnitude of the number fluctuation
mode (indicated by the arrows in figure 4(b))is much smaller in
the case of the interacting sample A. This observed suppression of
the number fluctuations has a simple qualitative explanation: a
particle trying to enter the scattering volume from one side will,

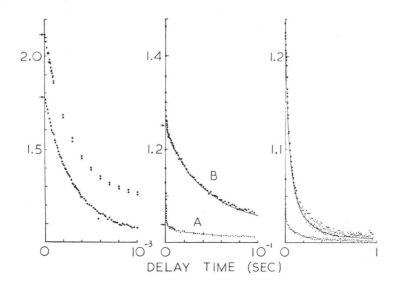

DELAY TIME (SEC)

Fig. 4.   Correlation functions for samples A and B.   Note changing
          vertical scales and wide range delay times spanned.   The
          decay in (a) is due almost entirely to interference
          fluctuations whereas the decays in (b) and (c) are due to
          number fluctuations whose magnitudes are indicated by the
          arrows in (b).   (Taken from reference 25).

due to the repulsive forces, tend to "push" a neighbouring
particle out of the other side thereby maintaining the number in
V more or less constant.

3.2   Theory and Interpretation (18,25)

     The theoretical treatment of this experiment is complicated
because diffraction smears the illumination profile of the small
scattering volume.   In fact, to make progress, it is necessary to
assume this profile to be "three-dimensional Gaussian", that is
$E(r) = \exp(-r^2/s^2)$ where r is measured from the centre of the
volume and s is an appropriate radius (5,7).   Here it is sufficient
for our purposes to outline a simpler theory which, in the main,
assumes a uniformly illuminated scattering volume.   Also, for
simplicity, we will neglect the interference fluctuations.   The
complete theoretical treatment is published elsewhere (25).

     The photon correlation measurements were made at scattering
angle $\theta = 90°$, well above the peak in the structure factor (figure
3; $\theta = 90°$ corresponds to $K = 1.87 \times 10^5$ cm^{-1} where $S(K) \approx 1$).   Thus
the scattered intensity I(t) can, when time-averaged over the rapid

interference fluctuations, be taken to be simply proportional to the instantaneous number $N(t)$ of particles in V (25), so that

$$\frac{<I(0)\ I(\tau)>}{<I>^2} = \frac{<N(0)\ N(\tau)>}{<N>^2} = 1 + \frac{<\Delta N(0)\ \Delta N(\tau)>}{<N>^2} \tag{16}$$

where the number fluctuation $\Delta N(t) \equiv N(t) - <N>$. Thus a measurement of the intensity correlation function extrapolated to $\tau = 0$ (but ignoring interference fluctuations) gives a direct estimate of number fluctuations $<\Delta N^2>/<N>^2$.

The magnitude of the number fluctuations can be related to the interaction through the static pair correlation function $g(r)$ as was first shown by Ornstein and Zernike in 1914 (26,27):

$$\frac{<\Delta N^2>}{<N>} = 1 + \frac{<N>}{V^2} \int_V d^3r_1 \int_V d^3r_2 \left[g(|\underline{r}_1 - \underline{r}_2|) - 1\right]. \tag{17}$$

For a three-dimensional Gaussian scattering volume of radius s equation (17) can be written, with use of the Fourier relationship between $g(r)$ and $S(K)$, as

$$\frac{<\Delta N^2>}{<N>} = \frac{s^3}{2\sqrt{\pi}} \int_0^\infty K^2 dK \exp(-K^2s^2/4)\ S(K), \tag{18}$$

a result first derived by Berne (18), which is the appropriate form of the Ornstein–Zernike expression for the analysis of our data. We note that for, $S(K) = 1$ (non-interacting particles), equation (18) gives

$$<\Delta N^2> = <N>, \tag{19}$$

the well-known result for a random or Poisson distribution of particles. For large enough V, i.e. $Ks \gg 1$, (18) reduces to another well-known expression (27)

$$\frac{<\Delta N^2>}{<N>} = S(0). \tag{20}$$

For the non-interacting sample B, for which $<I_B^2>/<I_B>^2 = 1.25$ (figure 4b), we obtain from (16) $<\Delta N_B^2>/<N_B>^2 = 0.25$ so that, from (19), $<N_B> = 4.0$. The mean occupation number $<N_A>$ of sample A can then be found by recognizing that the ratio, ~ 1.47, of intensities scattered at large K by the two samples (figure 3) is equal to the concentration ratio. Thus $<N_A> = 1.47 \times 4.0 = 5.88$. From figure 4b we have $<\Delta N_A^2>/<N_A>^2 = 0.037$ so that $<\Delta N_A^2>/<N_A> = 0.218$. This must be compared with 0.182, the value of the right-hand-side of equation (18), obtained by numerical integration of the data in figure 3.

These two values differ by about 20% whereas the estimated experimental error on each quantity was about 10%. This experiment therefore provides a rough verification of the Ornstein-Zernike expression, equation (17).

We note that, as in section 2.1, we have proceeded here as if the particles in question were monodisperse. In fact they are not and the relative standard deviation in particle size quoted by Dow is σ = 0.064. Analysis of the data allowing for polydispersity would require a detailed knowledge of the particle size distribution (for example, its 12'th moment, Appendix 2). An approximate correction is outlined in Appendix 2. Again we find a surprisingly large effect of quite a small degree of polydispersity. The magnitude of the correction is strongly dependent on the value taken for σ; σ = 0.064 gives a somewhat larger difference (of the opposite sign) between the corrected values of the two quantities measured than was found for the uncorrected values (Appendix 2).

3.3 Discussion

The work described in section 3 provides what appears to be the first direct experimental test of the 65-year-old Ornstein-Zernike expression (equation 17). While such tests of fundamental relationships are obviously important, the Ornstein-Zernike result is well established by less direct means and is unlikely to be found wanting. Other relevance of this work is not immediately obvious since, although the experiment is performed in the non-Gaussian regime, the quantity measured depends only on the pair correlation function (equation 17) which can, in general, be determined by simpler methods. However, for relatively large scattering volumes (K_{max} s>>1), the measurement is sensitive to the low-K behaviour of S(K) (equation 20) which, for various reasons, can be difficult to obtain in direct measurements of the angular dependence of the average intensity. It has therefore been suggested (25) that number-fluctuation measurements could be used to determine S(0) and hence osmotic pressures in very dilute suspensions of colloids, particularly those showing reversible aggregation.

On a more fundamental level, a simple extension of the experiment should provide higher (third - and fourth-order) analogues of S(K) and hence g(r) which, although they play an important role in the theory of liquids, are difficult to measure by other means (25). This extension involves making measurements at smaller K ($K \leqslant K_{max}$), where S(K) $\neq$ 1 and it is no longer permissible to take I(t) $\propto$ N(t) as in section 3.2.

4. NON-GAUSSIAN SCATTERING BY NON-INTERACTING NON-SPHERICAL PARTICLES (28)

4.1 Introduction

The optical arrangement for this experiment was similar to that described in section 3.1 except that a second detector was placed at 90° on the opposite side of the laser beam to the first (see figure 5). Both detectors observed the same small scattering volume (about $6\mu m^3$ in this case) which contained on average one particle. The outputs of the two detectors were cross-correlated. The sample was a dilute suspension of tobacco mosaic virus (TMV) a rod-shaped particle of length L ≈ 300nm and diameter about 15nm (negligible compared to λ). Enough electrolyte was present to shield interparticle interactions effectively.

Consider first the scattered intensity at one of the detectors. It will fluctuate in time due to three distinct mechanisms. Firstly there will be the usual interference fluctuations which depend on both translational and rotational motions. Secondly, there will be number fluctuations as discussed in section 3. Finally, because the particles are not spherical there will be non-Gaussian fluctuations in the scattered intensity associated with orientational motions; (these fluctuations would occur even if the (small) number of particles in V was fixed). Turning now to the cross-correlation between the signals from two detectors well-separated spatially, the interference fluctuations will not contribute to the correlation function because they are only correlated within one coherence area. The effect of the number fluctuations will be unaltered since each detector observes the same particles. However the non-Gaussian term associated with orientational motion can now become quite complicated because the light scattered by a single non-spherical particle into a given direction can be a strong function of its orientation with respect to that direction (equation 24). It is the purpose of this experiment to exploit this last fact.

4.2 Theory (28)

With several assumptions, the most important being (i) that the illuminating intensity does not vary significantly over the length of the rod (i.e. s >> L) and (ii) that rotational and translational motions of the rods are not coupled, it can be shown that the intensity cross-correlation function is:

$$\frac{<I(\underline{K}_1,0)\ I(\underline{K}_2,\tau)>}{<I(\underline{K}_1)>\ <I(\underline{K}_2)>} = 1 + g^T_{NG}(\tau)\ g^R_{NG}(\underline{K}_1,\underline{K}_2,\tau). \qquad (21)$$

This expression has the form expected from the discussion in section 4.1: $g_{NG}^T(\tau)$ is the translational or number fluctuation part of the non-Gaussian term (the second term in (16)) which, for non-interacting particles in a three-dimensional Gaussian scattering volume is (5,6)

$$g_{NG}^T(\tau) \; = \; <N>^{-1} \left[1 + (4 \; D_T \; \tau/s^2) \right]^{-3/2} \; . \tag{22}$$

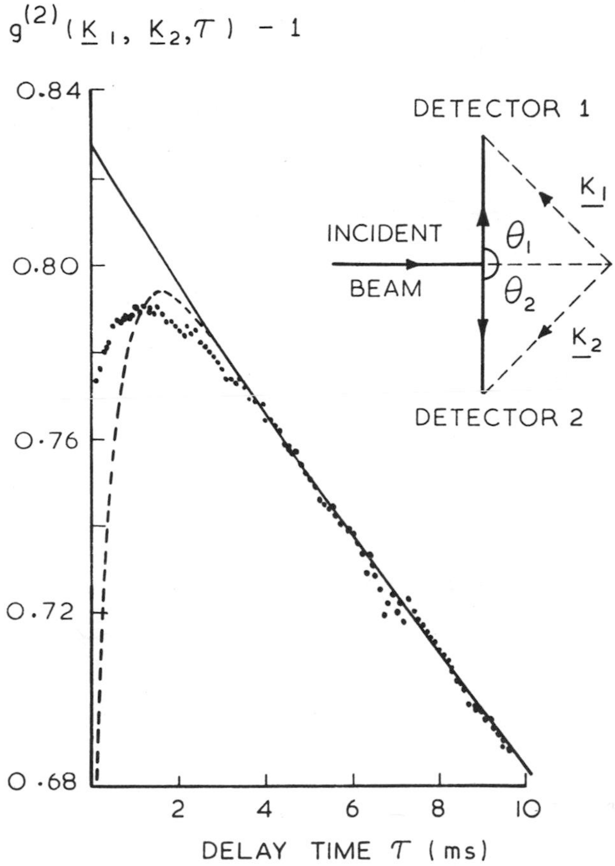

Fig. 5.   Lower left: Cross-correlation functions for scattering by TMV particles: points – experimental results; solid line – g_{NG}^T, translational (number fluctuation) part of correlation function; dotted line – full theoretical prediction. (Taken from reference 28).

Upper right: Plan view of deployment of detectors with relevant scattering angles and vectors.

(In section 4 only we denote the translational diffusion coefficent
by D_T and the rotational diffusion coefficient by D_R.) The
rotational contribution g_{NG}^R is

$$g_{NG}^R(\underline{K}_1,\underline{K}_2,\tau) = <a^2(\underline{K}_1,0)a^2(\underline{K}_2,\tau)>/<a^2(\underline{K}_1)> \; <a^2(\underline{K}_2)> \qquad (23)$$

where $a(\underline{K},t)$ is the amplitude of the light scattered by the TMV rod
at time $\bar{t}$. In terms of $\phi(t)$, the instantaneous angle between
$\underline{K}$ and the long axis of rod, $a(\underline{K},t)$ can be written:

$$a(\underline{K},t) = \frac{1}{L}\int_{-\frac{L}{2}}^{\frac{L}{2}} \exp\left[iKx \cos\phi(t)\right] dx. \qquad (24)$$

With use of the usual model for rotational Brownian motion, (23)
can be evaluated, with some difficulty, to give

$$<a^2(\underline{K}_1,0)a^2(\underline{K}_2,\tau)> = \sum_{\substack{\ell \\ even}} (2\ell+1)\; P_\ell(\cos\Phi)\; \exp\left[-\ell(\ell+1)D_R\tau\right] X$$

$$\sum_{\substack{p,p' \\ even}} \sum_{q,q'} (i)^{p+q-p'-q'} (2p+1)(2p'+1)(2q+1)(2q'+1)\; I_p(K_1)I_{p'}(K_1)I_q(K_2)$$

$$I_{q'}(K_2)X \begin{pmatrix} \ell & p & p' \\ 0 & 0 & 0 \end{pmatrix}^2 \begin{pmatrix} \ell & q & q' \\ 0 & 0 & 0 \end{pmatrix}^2 , \qquad (25)$$

where

$$I_p(K) = \frac{2}{KL} \int_0^{\frac{KL}{2}} j_p(y)dy, \qquad (26)$$

P_ℓ is a Legendre polynomial, j_p is a spherical Bessel function, Φ
is the angle between $\underline{K}_1$ and $\underline{K}_2$ and the relevant 3-j symbols are
tabulated in reference 29.

4.3 Results (28)

The solid line in figure 5 is the theoretical prediction for
the number fluctuation term g_{NG}^T, calculated from equation (22)
with appropriate values of D_T and s. As expected this is a purely
decaying function. The dashed line in figure 5 is the full
prediction for the second term in (21) obtained by multiplying

g_{NG}^T by g_{NG}^R, calculated from (25) with appropriate values of L, D_R, ϕ etc. We see that g_{NG}^R starts <u>below</u> 1, implying an anti-correlation between the signals scattered in opposite directions by orientational fluctuations, and grows towards its asymptotic value of 1 (equation 23) at large τ. Formally this anticorrelation arises from the $\ell = 2$ terms in (25) (five of which were appreciable) which are negative because P_2 (cos 90) = -0.5. Physically it has a simple qualitative explanation: If, at some instant, a particle lies in the scattering plane and its long axis bisects the angle θ_1 between the beam and detector 1 (figure 5), there is virtually no phase shift in the light scattered by different parts of the particle into this detector which therefore receives a strong signal (equation 24 with $\phi = 90^o$). However, for a particle in this orientation, there is an o tical path difference of about one wavelength between the light scattered from either end of the particle into detector 2 which, due to destructive interference, registers a weak signal (equation 24 with $\phi = 0^o$). Clearly, then, strong anticorrelations exist for certain particle orientations and the theory shows that a small effect survives the full average over orientation as the particle executes its rotational Brownian motion.

The experimental measurement is shown by the data points in figure 5. The data clearly show the qualitative features predicted by the theory, namely an anticorrelation in g_{NG}^R, decaying in about 1 ms. However quantitative agreement was not found. The most likely explanation of the marked reduction, observed experimentally, of the amplitude of the anticorrelation is significant variation of the illuminating intensity over a single particle (s ≈ 1μm, compared with L = 0.3μm) i.e. assumption (i) of section 4.2 does not hold. This will prevent total destructive interference (on which the magnitude of the anticorrelation depends crucially) from occurring in the light scattered by a particle, whatever its orientation. Attempts to incorporate this effect into the theory lead to discouraging complications. A more promising approach would be to use a larger scattering volume (larger s). This would need more dilute samples, so that <N> in equation (22) does not become too large, and should be possible in a carefully designed experiment. This approach has the added advantage of giving a greater separation between the timescales of the rotational and translational contributions (since g_{NG}^T depends on s whereas g_{NG}^R does not).

4.4 <u>Discussion</u>

We see from the form of equations (25) and (26), that the dependence of g_{NG}^R on K_1 and K_2 (and hence Φ) is, for rod-shaped particles, a sensitive function of particle dimension L. Experiments of this type therefore constitute, in principle, a new method for characterizing particles in dilute suspension (30). However, even if the difficulties discussed in section 4.3 can be

resolved, it is improbable that this method will become routine
because, in the general case, inversion of the experimental data
to obtain parameters of particles of arbitrary size and shape
seems likely to be prohibitively difficult. (However this does
not preclude the route followed in other particle-characterization
methods based on light scattering, namely the comparison of
data with theoretical calculations for specific particle shapes
and sizes.)

Regardless of these reservations, this experiment has
demonstrated the existence of a rather unusual effect. The
ability to observe the motions of a single particle (or at least a
small number of particles) has a certain intuitive appeal and it is
possible that the principles illustrated here could be exploited
further. An obvious example is the study of interacting rods
where rotations should be hindered and there is even the possibility
of detecting coupling between rotational and translational motions
(31).

5.   CONCLUDING REMARKS

The experiments reviewed in section 2, combined with parallel
theoretical developments (3) and computer simulations (4), have
led to a fairly complete qualitative understanding of the dynamics
of dilute suspensions of strongly-interacting particles.
Nevertheless, because of two main experimental complications,
polydispersity and multiple scattering, we are still some way from
obtaining a body of "good" experimental data to provide detailed
verification of the theories. The relative effect of multiple
scattering (hardly mentioned here but discussed in detail elsewhere
(32-34,15) is, for small particles, proportional to the sixth
power of the particle radius. It is an unfortunate circumstance
that preparation of small colloidal spheres which are not
significantly polydisperse has yet to be achieved. Therefore
at present one has the choice of working with small polydisperse
particles, for which the effect of multiple scattering is small
(as in section 2), or more monodisperse larger particles where
multiple scattering can be significant. Neither the multiple
scattering nor the polydispersity corrections are trivial. The
need for small (<50 nm diameter) monodisperse particles, for which
both complications can be minimized simultaneously, is clear.

The experiments described in sections 3 and 4 represent attempts
to derive some of the extra information which is, in principle,
available when small numbers of scatterers are studied (i.e. in the
non-Gaussian regime). Neither experiment was completely successful
and neither, at present, adds much to our knowledge of the systems
investigated. Nevertheless the feasibility of such experiments
has been demonstrated and some promising future directions

identified (sections 3.3 and 4.4). Experiments using small
scattering volumes are difficult both because of the need for
highly stable optics and because interpretation of the data
frequently depends heavily on detailed knowledge of the scattering
geometry (the illumination profile, for example). By contrast,
in Gaussian (large-scattering-volume) experiments, the scattering
vector, which is relatively easy to define, provides the only
important length scale.

ACKNOWLEDGEMENTS

     I am very grateful to Dr. M. B. Weissman for sending me a
preprint of his work on the effect of polydispersity on light
scattering by interacting particles. The work described in
section 4 was performed in collaboration with Dr. W. G. Griffin.
I thank Dr. Griffin and Professor R. H. Ottewill for many
valuable discussions.

APPENDIX 1:   INITIAL DECAY OF MEASURED CORRELATION FUNCTION FOR
              INTERACTING POLYDISPERSE PARTICLES

Equation (10) can be written

$$F^M(K,\tau) = \frac{1}{N\langle a^2\rangle}\left\langle\sum_{i=1}^{N}\sum_{j=1}^{N} a_i a_j \exp iK\left[x_i(0) - x_j(\tau)\right]\right\rangle, \qquad (A1)$$

where

$$F^M(K,0) = S^M(K)$$

and we have taken K to be in the x-direction. If the $\{a_i\}$ are
assumed to be time-independent, we obtain the well-known result:

$$\frac{d^2 F^M(K,\tau)}{d\tau^2} = -\frac{K^2}{N\langle a^2\rangle}\left\langle\sum_{i=1}^{N}\sum_{j=1}^{N} \langle a_i a_j v_i(0)\, v_j(\tau)\exp iK\left[x_i(0)-x_j(\tau)\right]\right\rangle \qquad (A2)$$

where v_i is the x-component of the velocity of particle i and the
"stationarity condition" (27) has been used. Integration of (A2)
gives, as an exact result for a classical system of particles:

$$\frac{dF^M(K,\tau)}{d\tau} = -\frac{K^2}{N\langle a^2\rangle}\sum_{i}\sum_{j}\int_{0}^{\tau} dt\langle a_i a_j v_i(0)v_j(t)\exp iK\left[x_i(0)-x_j(t)\right]\rangle \qquad (A3)$$

We now specialize to times short enough that a particle only moves
a small fraction of both the interparticle spacing and K^{-1}, where,
following the discussion of section 2.1, the particle can be
regarded as an essentially free diffuser (provided hydrodynamic
interactions are neglected). In this limit, therefore, the
velocities $v_i(t)$ in (A3) can be replaced by $v_i^B(t)$, the "Brownian"
velocities which would obtain in the absence of interactions (16).
Then, since $v_i^B(0)$ is not correlated with $v_j^B(t)$, $x_i(0)$ or $x_j(0)$,
only the $i = j$ terms are non-zero in (A3) so that (16)

$$\frac{dF^M(K,\tau)}{d\tau}\bigg|_{\text{small }\tau} = -\frac{K^2}{N\langle a^2\rangle}\sum_{i=1}^{N} a_i^2 \int_{0}^{\tau}\langle v_i^B(0)\, v_i^B(t)\rangle dt. \qquad (A4)$$

Finally τ is taken to be large compared to the typical (rapid) fluctuation time of v^B (which can generally be done without violating the previous restriction on τ (16)). The upper limit of the integral can then be taken to ∞ and its value identified with D_{oi}, the free-particle diffusion coefficient of species i. Thus the initial slope of the normalized correlation function is $<D_o>K^2/S^M(K)$ where

$$<D_o> = \sum_{i=1}^{N} a_i^2 D_{oi} \Big/ \sum_{i=1}^{N} a_i^2 \quad . \tag{A5}$$

APPENDIX 2:   THE EFFECT OF POLYDISPERSITY ON NUMBER FLUCTUATION
              MEASUREMENTS

We limit the discussion to a uniformly-illuminated scattering volume V and measurement angle well above the structure in $S(K)$ and we ignore interference fluctuations.   Then the instantaneous intensity can be written (18):

$$I(t) = \sum_{i=1}^{N_T} a_i^2 b_i(t) \tag{A6}$$

where N_T is the (large) number of particles in the total sample volume V_T, a_i is, as before, the scattering amplitude and $b_i(t)$ is a counting variable (6,25)

$$b_i(t) = 1 \quad \text{if particle } i \text{ is in } V$$
$$\tag{A7}$$
$$= 0 \quad \text{otherwise.}$$

Thus

$$<I> = <N> <a^2> \tag{A8}$$

where $N(t)$ is the instantaneous number of particles in V:

$$N(t) = \sum_{i=1}^{N_T} b_i(t) \quad . \tag{A9}$$

The intensity correlation function at $\tau \to 0$ is

$$<I^2> = \sum_{i=1}^{N_T} \sum_{j=1}^{N_T} <a_i^2 a_j^2><b_i b_j> = <a^2>^2 <N^2> + \left[<a^4>-<a^2>^2\right] <N> \quad (A10)$$

where we have used (A9) and the fact that $b_i^2 = b_i$ (A7). Thus from (A8) and (A10) we have

$$\frac{<N^2>}{<N>^2} = \frac{<I^2>}{<I>^2} - \left[\frac{<a^4>}{<a^2>^2} - 1\right] \frac{1}{<N>} \quad (A11)$$

whereas (12) can be rewritten

$$S^I(K) = \frac{<a^2>}{<a>^2} \left[S^M(K) - 1\right] + 1. \quad (A12)$$

The correction of the measured quantity, $<I^2>/<I>^2$, to obtain the quantity desired, $<N^2>/<N>^2$, thus requires knowledge of $<a^4>$ and hence $<R^{12}>$ for small particles (A11). Here we can only use equation (14) though its validity for such large n is questionable for any but the most narrow particle size distributions. For non-interacting particles ($<\Delta N^2>=<N>$) equation (A11) becomes

$$\frac{1}{<N>} = \frac{<a^2>^2}{<a^4>} \left[\frac{<I^2>}{<I>^2} - 1\right] \quad (A13)$$

which can be used to obtain $<N>$ for sample B and hence sample A (by use of the intensity ratio, section 3.2). Equation (A11) can then be used to obtain the corrected value of $<\Delta N_A^2>/<N_A>$.

The correction for the right hand side of equation (18) is relatively small and, for simplicity, we will apply equation (A12) after the integration in (18) rather than before as we should strictly.

Corrected values for the two quantities measured are given in Table 2 for two values of standard deviation, $\sigma = 0.064$, the value quoted by Dow, and $\sigma = 0.05$. First we see that the correction for the number fluctuations is large even for quite small σ. Secondly we note that this admittedly approximate correction actually improves agreement between experiment and theory if we take $\sigma = 0.05$!

TABLE 2

Polydispersity correction to number fluctuation and
structure factor measurements. S'(0) is right-hand-
side of equation (18), σ is relative standard deviation
of particle size distribution.

	UNCORRECTED	CORRECTED	
		σ = 0.05	σ = 0.064
$\dfrac{<\Delta N_A^2>}{<N_A>}$	0.218	0.147	0.102
S'(0)	0.182	0.164	0.152

REFERENCES

1. M. B. Weissman, J. Chem. Phys. 72:231 (1980).
2. F. Grüner and W. Lehmann, these proceedings.
3. W. Hess, these proceedings.
4. K. J. Gaylor, I. K. Snook, W. van Megen and R. O. Watts, these proceedings.
5. For a general discussion of Gaussian and non-Gaussian scattering see P. N. Pusey, in "Photon Correlation Spectroscopy and Velociemetry", H. Z. Cummins and E. R. Pike, eds. Plenum, New York (1977).
6. D. W. Schaefer and B. J. Berne, Phys. Rev. Lett. 28:475 (1972)
7. D. W. Schaefer, Science 180:1293 (1973).
8. M. Weissman, H. Schindler and G. Feher, Proc. Nat. Acad. Sci. USA 73:2776 (1976).
9. J. C. Brown, P. N. Pusey, J. W. Goodwin and R. H. Ottewill, J. Phys. A (GB) 8:664 (1975).
10. P. S. Dalberg, A. Bøe, K. A. Strand and T. Sikkeland, J. Chem. Phys. 69:5473 (1978).
11. D. W. Schaefer, J. Chem. Phys. 66:3980 (1977).
12. R. Williams and R. S. Crandall, Phys. Lett. 48A:225 (1974).
13. D. W. Schaefer and B. J. Ackerson, Phys. Rev. Lett. 35:1448 (1975).
14. N. A. Clark, A. J. Hurd and B. J. Ackerson, Nature 281:57 (1979).
15. P. N. Pusey, J. Phys. A. (GB) 11:119 (1973).
16. P. N. Pusey, J. Phys. A. (GB) 8:1433 (1975).
17. B. J. Ackerson, J. Chem. Phys. 64:242 (1976).
18. B. J. Berne, in "Photon Correlation Spectoscopy and Velocimetry", H. Z. Cummins and E. R. Pike, eds., Plenum, New York (1977).
19. P. N. Pusey, Phil. Trans. Roy. Soc. Lond. A 293:429 (1979).
20. K. Alexander, unpublished data.
21. J. C. Brown, P. N. Pusey and R. Dietz, J. Chem. Phys., 62:1136 (1975).
22. D. J. Cebula, R. H. Ottewill, P. N. Pusey and J. Ralston, to be published.
23. G. Maisano, F. Mallamace and F. Wanderlingh, these proceedings.
24. P. A. Egelstaff ed., "Thermal Neutron Scattering,", Academic, London (1965).
25. P. N. Pusey, J. Phys. A. (GB) 12:1805 (1979).
26. L. S. Ornstein and F. Zernike, Proc. Akad. Sci. (Amsterdam) 17:793 (1914), reprinted in "The Quilibrium Theory of Classical Fluids", H. L. Frisch and J. L. Lebowitz, eds., Benjamin, New York (1964).
27. P. A. Egelstaff, "An Introduction to the Liquid State", Academic, London (1967).
28. W. G. Griffin and P. N. Pusey, Phys. Rev. Lett. 43:1100 (1979).
29. D. M. Brink and G. W. Satchler, "Angular Momentum", Oxford University Press (1968).

30. Z. Kam, Macromolecules 10:927 (1977).
31. W. G. Griffin, private communication.
32. A. Bøe and O. Lohne, Phys. Rev. A. 17:2023 (1978).
33. C. M. Sorensen, R. C. Mockler and W. J. O'Sullivan, Phys. Rev. A. 17:2030 (1978).
34. F. Grüner and W. Lehmann, Multiple scattering of light in a system of interacting Brownian particles, preprint.
35. D. Magde, E. Elson and W. W. Webb, Phys. Rev. Lett., 29:705 (1972).
36. W. W. Webb, Quart. Rev. Biophys. 9:49 (1976).

DIFFUSION COEFFICIENTS IN COLLOIDAL AND POLYMERIC SOLUTIONS

Walter Hess

Fakultät für Physik
Universität Konstanz
D-7750 Konstanz, Germany

INTRODUCTION

During the last years great efforts have been made to determine the diffusion coefficients of colloidal and polymeric solutions by means of quasielastic light scattering. To interpret and to understand these experiments several theories have been developed. Due to different starting points and the differences in the physical structure of the systems under consideration, these theoretical results often appear very different. Therefore the question arises whether it is possible to find a general expression for the diffusion coefficient of such systems and to relate the various other theories to it. The most general and rigorous formula for the diffusion coefficient has been given by Ackerson (1) and it is the main point of this paper to relate his result to other existing theories.

The experimental determination of a diffusion coefficient in different physical systems, e.g. suspensions of charged particles and polymeric solutions, is complicated by the fact that the measured time-dependent concentration autocorrelation function is not a simple exponential. The characteristic times of this function are strongly wave-vector dependent. Such a behaviour can be described by a memory equation. In section III some of the properties of the memory function will be discussed and it will be shown how information about it can be extracted from experiments.

Quasi-elastic light scattering measures a collective or mass diffusion coefficient, a quantity for which it is difficult to get an intuitive understanding. In constrast, the self-diffusion coefficient seems to be easier to understand. A very rough

31

description of some of its properties will be given in section IV.

THE COLLECTIVE OR MASS DIFFUSION COEFFICIENT

The statistical properties of colloidal or polymeric solutions can be completely described by the Smoluchowski equation

$$\{\frac{\partial}{\partial t} - \hat{L}^o(\{\underline{R}_\alpha\}) - \hat{L}^e(\{\underline{R}_\alpha\},t)\} \, P(\{\underline{R}_\alpha\},t) \; = \; 0, \tag{2.1}$$

where $P(t)$ is the distribution function for the coordinates of N Brownian particles and $\hat{L}^o$ and $\hat{L}^e(t)$ are the Smoluchowski operators for an equilibrium system and the additional term caused by external force, respectively,

$$\hat{L}^o(\{\underline{R}_\alpha\}) \; = \; \sum_{i,j} \frac{\partial}{\partial \underline{R}_i} \, \underline{\underline{D}}_{ij} \, (\{\underline{R}_\alpha\}) \, (\frac{\partial}{\partial \underline{R}_j} - \beta \cdot \underline{F}_j(\{\underline{R}_\alpha\})), \tag{2.2}$$

$$\hat{L}^e(\{\underline{R}_\alpha\},t) \; = \; - \sum_{i,j} \frac{\partial}{\partial \underline{R}_i} \, \underline{\underline{D}}_{ij}(\{\underline{R}_\alpha\}) \cdot \beta \cdot \underline{F}_j^{\,e}(\{\underline{R}_\alpha\},t \; . \tag{2.3}$$

The tensors $\underline{\underline{D}}_{ij}$ represent the hydrodynamic interaction between particles i and j, $\underline{\underline{D}}_{ij}$ will in general depend on the coordinates of all particles, and $\underline{F}_j(\{\underline{R}_\alpha\})$ is the force on particle j due to the interactions with the other Brownian particles,

$$\underline{F}_j(\{\underline{R}_\alpha\}) \; = \; - \frac{\partial}{\partial \underline{R}_j} \, U_N(\{\underline{R}_\alpha\}), \tag{2.4}$$

$U_N(\{\underline{R}_\alpha\})$ is the interaction potential. This will be a mean interaction with respect to the influence of the solvent particles. $\underline{F}_j^{\,e}(\{\underline{R}_\alpha\},t)$ gives the action of an external force, $\beta = (k_B T)^{-1}$.

In a colloidal solution the Brownian particles are the colloid particles themselves, whereas in a polymeric solution a Brownian particle will represent one statistical segment of a polymer.

The light scattering spectrum is related to the concentration autocorrelation function

$$C(\underline{k},t) \; = \; <\delta c(\underline{k},t) \, \delta c(-\underline{k},0)> \; , \tag{2.5}$$

$$\delta c(\underline{k},t) \; = \; \sum_i e^{-i \, \underline{k} \cdot \underline{R}_i(t)} - (2\pi)^3 \, c \cdot \delta(\underline{k}) \; , \tag{2.6}$$

c is the mean concentration N/V and

$$<(\ldots)> \equiv \int d\{\underline{R}_\alpha\}(\ldots)e^{-\beta U_N(\{\underline{R}_\alpha\}}/\int d\{\underline{R}_\alpha\}e^{-\beta U_N(\{\underline{R}_\alpha\})} \qquad (2.7)$$

denotes the equilibrium expectation value.

For $C(\underline{k},t)$ a memory equation can be derived, which may be written in the form

$$\frac{\partial}{\partial t} C(\underline{k},t) = - D(\underline{k}) \cdot \underline{k}^2 \cdot C(\underline{k},t) \qquad (2.8)$$

$$+ \frac{1}{S(\underline{k})} \int_0^t dt' M(\underline{k},t-t') C(\underline{k},t') \quad \text{for } t > 0$$

with

$$D(\underline{k}) = \frac{1}{S(\underline{k})} \frac{1}{N} \sum_{i,j=1} <\hat{\underline{k}} \underline{\underline{D}}_{ij}(\{\underline{R}_\alpha\}) \cdot \hat{\underline{k}} \cdot e^{-i\underline{k}(\underline{R}_i - \underline{R}_j)}> \qquad (2.9)$$

$S(\underline{k})$ is the static structure factor, $S(\underline{k}) = \frac{1}{N} \cdot C(\underline{k},0)$, and $M(\underline{k},t)$ is the memory function, $\hat{k} = \underline{k}/|\underline{k}|$.

An exact statistical mechanical expression for the memory-function can be given, but this is very complicated and gives no physical insight into its meaning. We will delay a discussion of the memory-term in eq. (2.8) to the next section. Here we need only the property that it contributes to eq. (2.8) in the limit of small wave vectors only by three-particle correlations. Neglecting these gives

$$\lim_{k\to 0} M(\underline{k},t) / \underline{k}^2 = 0 \qquad (2.10)$$

Eqs. (2.8 - 2.10) have been rigorously derived by Ackerson (1) and Dieterich and Peschel(2) using the Mori-Zwanzig projection operator formalism, the latter paper neglect the hydrodynamic interaction for simplicity. Another derivation can be performed by using a nonlinear local equilibrium expansion of the configuration distribution function $P(\{\underline{R}_\alpha\},t)$, which leads to the same results (3,4).

Expanding eq. (2.8) to lowest order in $\underline{k}$ gives (provided that $S(0) \neq 0$), because of eq. (2.10), an ordinary diffusion-equation

$$\frac{\partial}{\partial t} C(\underline{k},t) = - D_c \cdot \underline{k}^2 \cdot C(\underline{k},t) \quad \text{for } t > 0, \qquad (2.11)$$

with the collective or mass diffusion coefficient

$$D_c = \frac{1}{N} \sum_{i,j=1}^{N} <\hat{k} \cdot \underline{\underline{D}}_{ij}(\{\underline{R}_\alpha\}) \cdot \hat{k}> / S(0), \qquad (2.12)$$

This microscopic expression for the diffusion coefficient seems to have been first proposed by Altenberger (5) by heuristic arguments. Using the continuity equation one can further easily show that D_c fulfills a Kubo relation,

$$D_c = \frac{1}{S(0)} \cdot \frac{1}{3} \int_0^\infty dt \frac{1}{N} \sum_{i,j} <\underline{V}_i(t) \cdot \underline{V}_j(0)> , \qquad (2.13)$$

$\underline{V}_i(t)$ is the velocity of particle i.

On the other hand, since the time-integral over the velocity correlation function is just the microscopic expression for the inverse of the drag coefficient, which may be seen by starting from the Fokker-Planck equation and using linear-response theory, and since the static structure faction at $\underline{k} = 0$ is related to the isothermal osmotic compressibility,

$$S(0) = \beta / \frac{\partial c}{\partial \pi} \Big|_T , \qquad (2.14)$$

D_c is equivalent to the Stokes-Einstein expression of irreversible thermodynamics for the diffusion coefficient

$$D_c = \frac{1}{f} \cdot \frac{\partial \pi}{\partial c} \Big|_T . \qquad (2.15)$$

It is important to note that the quantities in this relation are defined in a solvent-fixed reference frame. But drag coefficients are usually measured by non-equilibrium experiments like a sedimentation experiment by which a macroscopic solvent current is produced. Therefore a volume-fixed drag coefficient is measured, and to insert it into eq. (2.13) a reference frame correction has to be added (6,7). We have seen so far that eq. (2.12) is the correct microscopic expression for the diffusion-coefficient as it is measured by light scattering. It relates the diffusion coefficient in a general way to the equilibrium distribution function $\exp(-\beta U_N(\{\underline{R}_\alpha\}))$ and to the hydrodynamic interaction. In the literature there are many forms of D_c which look quite different from eq. (2.12) since they treat specific physical systems with different equilibrium distribution functions and use different approximations for the hydrodynamic interaction. In the following I will discuss some of these papers and distinguish between such dealing with systems with short-range forces, Coulomb forces and polymeric solutions.

Systems with Short-Range Forces

For low-concentrated systems with short-range forces the hydrodynamic interaction can be written as two-particle interaction, see e.g. ref. 8,

$$\underline{\underline{D}}_{ij}(\{\underline{R}_\alpha\}) = D_o(\underline{1} + \sum_k \underline{\underline{A}}(\underline{R}_i - \underline{R}_k))\delta_{ij} + D_o \cdot \underline{\underline{B}}(\underline{R}_i - \underline{R}_j)(1-\delta_{ij}), \quad (2.16)$$

D_o is the diffusion coefficient of a particle without interaction, $\underline{\underline{A}}$ gives the change of the mobility due to the presence of other particles and $\underline{\underline{B}}$ gives the velocity of particle i due to a force acting on <u>another</u> particle j.

A virial expansion of the equilibrium values in eq. (2.12) can be performed,

$$< \hat{k} \cdot \underline{\underline{D}}(\{\underline{R}_\alpha\}) \cdot \hat{k}> = \quad (2.17)$$

$$= D_o\{1 + c \int d^3r \ (\hat{k} \cdot \underline{\underline{A}}(\underline{r}) \cdot \hat{k} + \hat{k} \cdot \underline{\underline{B}}(\underline{r}) \cdot \hat{k}) \ e^{-\beta U_2(\underline{r})}\} ,$$

$$S(0) = 1 - c \int d^3r \ (1 - e^{-\beta U_2(\underline{r})}) , \quad (2.18)$$

$U_2(\underline{r})$ is the two-particle potential.

To first order in c we get finally

$$D_c = D_o(1 + (2 \cdot A_2 \cdot M - k_s) \cdot c) \quad (2.19)$$

in the notation of Pyun and Fixman (9) with

$$2 \cdot A_2 \cdot M = \int d^3r \ (1 - e^{-\beta U_2(r)}) \quad (2.20)$$

and

$$k_s = - \int d^3r \ \hat{k}(\underline{\underline{A}}(r) + \underline{\underline{B}}(\underline{r}))\hat{k} \ e^{-\beta U_2(r)} . \quad (2.21)$$

The mostly used approximation for the hydrodynamic interaction is the Oseen tensor, $\underline{\underline{A}}(\underline{r}) = 0$ and

$$\underline{\underline{B}}(\underline{r}) = \frac{3}{4} D_o \frac{a}{|\underline{r}|} \{\underline{1} + \hat{r} : \hat{r}\} \quad (2.22)$$

which has the important property

$$\frac{\partial}{\partial \underline{r}} \underline{\underline{B}}(\underline{r}) = 0 \quad (2.23)$$

Using now a hard-core potential and eqs. (2.22) and (2.23) one
gets directly the result of Altenberger and Deutch (10),

$$D_c = D_o(1 + 2\phi) , \qquad (2.24)$$

ϕ is the volume concentration. *

The Oseen tensor is certainly not a good approximation for the
hydrodynamic interaction, especially in a system with short-range
potential, since it is valid for large distances. There exist
significant improvements by works of Batchelor (12) and
Felderhof (13), who both also calculated D_c for a hard-core
system (8,15), Batchelor:

$$D_c = D_o(1 + 1.45\phi) \qquad (2.25)$$

Felderhoff:

$$D_c = D_o(1 + 1.56\phi) \qquad (2.26)$$

Besides the hard-core model there exist also calculations for a
square-well potential by Altenberger, using the Oseen tensor (5)
and an improved Oseen-tensor (16).

Phillies (7) took into account a screened Coulomb potential
besides the hard core. His approach is semimicroscopic in the
sense that an empirical expression for the friction coefficient
is used but a microscopic one for the osmotic pressure term,

$$D_c = \frac{1}{f} \{\beta^{-1} - c \frac{4\pi}{3} \int_0^\infty dr \, r^3 \frac{\partial U_2(r)}{\partial r} g(r)\} , \qquad (2.27)$$

$g(r)$ is the radial distribution function. This result is only to
first order in c equivalent to eq. (2.12). Phillies also derives
the Stokes-Einstein relation (2.15), but again this derivation is
only valid to first order in c, although the general result is
obtained.

Systems with Coulomb Forces

An important contribution to the understanding of the diffusion
coefficient for systems with long ranged electrostatic forces has
been given by Stephen (17). The hydrodynamic interaction is

*An attempt by Hess and Klein (11), to improve the result of
Altenberger and Deutch by taking into account memory-function effects
is erroneous since eq. (2.10) was not satisfied.

completely neglected and the starting point is the linearized
Poisson-Boltzmann equation. The result can be written as

$$D_c = \frac{D_o}{S(0)} \tag{2.28}$$

and $S(0)$ has the Debye-Hückel form

$$S(0) = \frac{\sum\limits_i q_i^2 \, c_i}{Q^2 c + \sum\limits_i q_i^2 \cdot c_i} \; , \tag{2.29}$$

Q is the charge of the Brownian particles and q_i and c_i are the
charges and concentrations, respectively, of the different kinds
of ions in the solvent. The theory is also valid when the ionic
strength is small so that only the counterions are present. Using
the electro-neutrality condition one gets

$$S(0) = |q/Q| \; , \tag{2.30}$$

q is the charge of the counterions, and

$$D_c = D_o \cdot \left| \frac{Q}{q} \right| \tag{2.31}$$

is independent of the concentration, due to the long range Coulomb
forces.

Doherty and Benedek (18) have measured diffusion coefficients in
systems of charged Bovin Serum Albumin but they found significant
deviations from Stephen's theory. They plotted their results as
$\alpha \equiv (D_c - D_o)/D_o$ which turns out to be always much smaller than the
theoretical value

$$\alpha = \frac{Q^2 c}{\sum\limits_i q_i^2 \, c_i} \tag{2.32}$$

The deviation is especially pronounced at low ionic strength.
They believe that these deviations are due to the fact that
their systems are too concentrated, so that a Debye-Hückel
approximation can not be applied. But it may be that also the
neglect of hydrodynamic interaction is partially responsible for
the deviations. If we use as the simplest possible approximation
the Oseen tensor (2.22) for the hydrodynamic interaction we get
to lowest order in the ionic strength

$$\alpha = \frac{Q^2 \, c}{\sum\limits_i q_i^2 \, c_i} \; (1 - \frac{a}{\lambda}) \; , \tag{2.33}$$

λ is the Debye-Hückel screening length. With hydrodynamic interaction α is always smaller than without, and since in Doherty and Benedeks systems a $\simeq$ 35 Å and λ is of the order of 100 Å the hydrodynamic interaction gives a correction of about 37%, which is of the order of magnitude of the deviations between the experimental α and Stephen's α.

Schor and Serrallach (19) have tried to apply the theory of Phillies to the experiments of Doherty and Benedek. It seems to be doubtful whether this approach is here possible, since Phillies theory is assumed to be valid for a short-range interaction. Anyway it could be substantially improved by taking into account hydrodynamic interaction.

Polymeric Solutions

Polymeric solutions are something specific since here the Brownian particles are bound to their next neighbours and n such Brownian particles of statistical segments form a chain polymer. The two-particle radial distribution function is therefore not only a function of the distance between the particles but depends explicitly on the particle indices, e.g. One has to distinguish the dilute solutions from the semidilute ones. In dilute solutions the polymers are far separated and only the diffusion coefficient of a single polymer is measured. In semidilute solutions the polymers are overlapping and form a fluctuating network. In this regime the diffusion coefficient is concentration dependent (20,21). Using general scaling arguments de Gennes (22,23) has predicted the dependence of the diffusion coefficient on the thermodynamic and interaction parameters in these different regimes.

Akcasu and coworkers used a formula like eq. (2.12), which they also derived by the projection-operator method from the Smoluchowski equation, although they did not show that the memory-function does not contribute to the diffusion coefficient (24-26). For the hydrodynamic interaction they use the Oseen-tensor, for the two-particle distribution function a Gaussian. The halfwidth of the Gaussian is given by the mean distance square between two particles, $<(\underline{R}_i - \underline{R}_j)^2>$. For these parameters a heuristic ansatz is used, which becomes exact at the θ-point. The result for D_c becomes especially simple for dilute solutions since here $S(0) = n$, and

$$D_c = \frac{1}{n^2} \sum_{i,j=1}^{n} < \hat{k} \, \underline{\underline{D}}_{ij}(\{\underline{R}_\alpha\}) \cdot \hat{k} > \qquad (2.34)$$

If one inserts the preaveraged Oseen tensor(2.22) one gets Kirkwood's diffusion coefficient (27),

$$D_c = \frac{D_o}{n} \{1 + \frac{1}{n} \sum_{\substack{i,j \\ i \neq j}} < \frac{a}{|\underline{R}_i - \underline{R}_j|} >\} , \qquad (2.35)$$

D_o is here the diffusion coefficient of a single segment. For long polymers the second term becomes more important than the first one and in the dilute regime the diffusion coefficient can be described by the Stokes formula

$$D_c = \frac{k_B T}{6 \pi \eta R_H} , \qquad (2.36)$$

η is the solvent viscosity and R_H a hydrodynamic radius, which is proportional to the radius of gyration of a polymer. R_H contains the dependence of the diffusion coefficient on excluded volume parameter and the length of the polymer (24,26). The theoretical results are in good agreement with experiments (20,21) and scaling predictions (22,23).

MEMORY EFFECTS

Light scattering experiments measure the hydrodynamic diffusion coefficient (2.12) only if the scattering vector is much smaller than the inverse of the correlation length. The correlation length is in weakly coupled systems of charged particles given by the Debye-Hückel length,in strongly coupled systems, where the Debye-Hückel length is shorter than the mean two-particle distance, it will be of the order of this mean two-particle distance. In dilute polymeric solutions the correlation length is the mean end to end distance of the polymers, in semidilute solutions it is characterized by the distance between two entaglements of different polymers. Many experiments, especially on systems of charged particles (28-33) and polymeric solutions (34-36) are performed at scattering vectors much larger then the inverse of the correlation length. In these experiments one has found that the correlation function shows significant deviations from an exponential behaviour and that the characteristic decay times are strongly wavevector dependent.

This non-exponential behaviour cannot be described by a diffusion equation like eq. (2.11), even if one would use a k-dependent generalized diffusion coefficient $D(\underline{k})$, eq. (2.9).

It is well known (37,38) that $D(\underline{k})$ gives the first cumulant of the correlation function, therefore describing the short time behaviour, but the long time behaviour is related to the memory function. This can easily be seen if we make a cumulant ansatz

$$C(\underline{k},t) = C(\underline{k},0) \exp \{ - \kappa_1(\underline{k}) \cdot t - \frac{1}{2} \kappa_2(\underline{k}) t^2 \ldots \} \tag{3.1}$$

in the memory-equation (2.8). We find for the first two cumulants

$$\kappa_1(\underline{k}) = D(\underline{k}) \cdot \underline{k}^2 \tag{3.2}$$

and

$$\kappa_2(\underline{k}) = \frac{M(\underline{k},0)}{S(\underline{k})} \quad ; \tag{3.3}$$

the higher cumulants can also be related to the memory function, but since nothing is known about the convergence of this expansion, it seems more convenient to characterize the long time behaviour by defining a mean relaxation time $\tau_m(\underline{k})$ as

$$\tau_m(\underline{k}) = \frac{\int_0^\infty dt\ C(\underline{k},t)}{C(\underline{k},0)} = \frac{\tilde{C}(\underline{k},0)}{\int_{-\infty}^\infty dz\ \tilde{C}(\underline{k},z)} = \frac{\tilde{C}(\underline{k},z=0)}{C(\underline{k},t=0)} , \tag{3.4}$$

where $\tilde{C}(\underline{k},z)$ is the Laplace transform of $C(\underline{k},t)$,

$$\tilde{C}(\underline{k},z) = \int_0^\infty dt\ e^{-zt} C(\underline{k},t) \quad . \tag{3.5}$$

Laplace transforming the memory-equation we get the solution

$$\tilde{C}(\underline{k},z) = \frac{N \cdot S(\underline{k})}{z + D(\underline{k})\underline{k}^2 - \dfrac{\tilde{M}(k,z)}{S(\underline{k})}}$$

and the inverse of $\tau_m(\underline{k})$ is

$$\tau_m^{-1}(\underline{k}) = D(\underline{k}) \cdot \underline{k}^2 - \tilde{M}(\underline{k},0)/S(\underline{k}) \quad . \tag{3.7}$$

$M(\underline{k},0)$ is a measure for the difference between first cumulant and τ_m^{-1} . If the correlation function is a sum of exponentials,

$$C(\underline{k},t) = \sum_i \alpha_i \exp - |t| /\tau_i \quad , \tag{3.8}$$

$$\tau_m = \sum_i \alpha_i \tau_i \quad , \tag{3.9}$$

and τ_m gives indeed the weighed mean of the relaxation times.

The microscopic expression of $M(\underline{k},t)$, which is rigorous within the applicability of eq. (2.1) is (1,2)

$$M(\underline{k},t) = <F(\underline{k},\{\underline{R}_\alpha\})\exp\ \{(1-\hat{P})\hat{L}_o(\{\underline{R}_\alpha\})t\}\cdot F(-\underline{k},\{\underline{R}_\alpha\})> \qquad (3.10)$$

with the generalized "random forces"

$$F(\underline{k},\{\underline{R}_\alpha\}) = (1-\hat{P})\hat{L}_o^+(\{\underline{R}_\alpha\})\cdot \delta c(\underline{k},0)\ . \qquad (3.11)$$

$\hat{L}_o^+(\{\underline{R}_\alpha\})$ is the hermitean adjoint operator of the Smoluchowski operator (2.2) and $\hat{P}$ is a projection operator which projects an arbitrary function of the particle configurations on the concentration fluctuations,

$$\hat{P}\ A(\{\underline{R}_\alpha\}) = \frac{<A(\{\underline{R}_\alpha\})\cdot \delta c(-\underline{k},0)>}{N\cdot S(\underline{k})}\cdot \delta c(\underline{k},0)\ . \qquad (3.12)$$

Dieterich and Peschel have calculated eq. (3.10) for a one-dimensional hard-core system without hydrodynamic interaction in the low density regime, with the result (2)

$$M(\underline{k},t) = 2c\ D_o\ k^2\ \sin^2 ka\cdot \left[\frac{2D_o}{\pi\cdot t}\right]^{1/2} \exp\left[-\frac{D_o k^2}{2}\right]t \qquad (3.13)$$

One finds that a cumulant expansion diverges since $M(\underline{k},0) = \infty$, the short-time expansion of $C(\underline{k},t)$ is of the type

$$C(\underline{k},t) = C(\underline{k},0)\ \exp\left[-\kappa_1(\underline{k})\cdot t + 0(t^{3/2})\right] \qquad (3.14)$$

Nevertheless, the integral over $M(\underline{k},t)$ exists and the mean relaxation time $\tau_m(\underline{k})$ turns out to be

$$\tau_m^{-1}(\underline{k}) = D_o\ \underline{k}^2\ (1 + 2\frac{c}{k}\sin 2\ ka)- D_o\ \underline{k}^2\ 4\frac{c}{k}\sin^2 ka\ . \qquad (3.15)$$

The second term results from the memory-function. At large wavevectors it becomes of the same magnitude as the concentration dependent part of the first term. Due to this memory-term $\tau_m(\underline{k})$ is larger than the inverse of the first cumulant. The correlation function decays at long times slower than at short times, which can be understood by considering the memory-equation and which is in accordance with the experimental findings (28-36).

The physical processes and correlations which are contained in eq. (3.10) are very complicated and therefore the insights into the meaning of the memory-function in this general formulation is very

limited. Moreover, a calculation of (3.10), which goes beyond such
a special model like the one mentioned above seems to be a
difficult task. We therefore need a simplified version of eq.
(3.10). As a very powerful method to obtain this the mode-mode
coupling theory has been proven. The macroscopic dynamic
behaviour of a system is largely determined by the fluctuations
of the conserved quantities, since these are the variables with
the longest relaxation times. In a system of Brownian particles
the only conserved quantity is the number of particles and
the corresponding fluctuations are the concentration fluctuations
The mode-mode coupling theory assumes that products of fluctuations
of conserved variables are the next important ones, since
although not conserved, they will also decay on a slow timescale
(40). We get therefore the lowest order mode-mode coupling approximation if we assume that besides the concentration fluctuations
also their products have to be considered as relevant variables.
This means that we approximate the operator $\{1-P\}$ in eqs. (3.10) and
(3.11) by the projection operators which project upon all bilinear
products $\delta c(\underline{k},0) \cdot \delta c(\underline{k}',0)$. Then the memory-function becomes (39)

$$M(\underline{k},t) = \frac{1}{2} \cdot c \cdot \frac{1}{(2\pi)^3} \int d^3k' \; g(\underline{k},\underline{k}') \cdot C(\tfrac{1}{2}\underline{k}+\underline{k}',t) \cdot \qquad (3.16)$$

$$\cdot C(\tfrac{1}{2}\underline{k}-\underline{k}',t) \cdot g(\underline{k},\underline{k}') \; .$$

Neglecting for simplicity hydrodynamic interactions and making a
Gaussian approximation for the equal-time correlation functions
we get

$$g(\underline{k}.\underline{k}') = D_o \underline{k}(\underline{k} - \tfrac{1}{2} \underline{k}') \; h_D(\underline{k} - \tfrac{1}{2}\underline{k}') \; / \; S(\underline{k} + \tfrac{1}{2} \underline{k}') \qquad (3.17)$$

$$+ D_o \underline{k}(\underline{k} + \tfrac{1}{2} \underline{k}') \; h_D(\underline{k} + \tfrac{1}{2}\underline{k}') \; / \; S(\underline{k} - \tfrac{1}{2} \underline{k}') \; ,$$

where $h_D(\underline{k})$ is the direct correlation function,

$$h_D(\underline{k}) = \frac{1 - S(k)}{c \cdot S(k)} \qquad\qquad (3.18)$$

It follows that $\lim_{k\to 0} g(\underline{k},\underline{k}')/|\underline{k}| = 0$, therefore eq. (2.10) holds
also in mode-mode coupling approximation. The result (3.16) can
be interpreted in the following way. Let us suppose that at some
initial time there is a sinusoidal concentration distribution with
wavevector $\underline{k}$. Due to the thermal forces of the solvent and the
interactions between the particles this distribution will become
smeared out and its amplitude decays, which is described by the
$D(k) \cdot k^2$ term in eq. (2.8). But because of the interactions also
new fluctuations with wavevectors $\tfrac{1}{2} \underline{k} + \underline{k}'$ and $\tfrac{1}{2} \underline{k} - \underline{k}'$ will be

created, the probability for this process being proportional to
the vertex $g(\underline{k},\underline{k}')$. These concentration fluctuations decay also
with their propagators $C(\frac{1}{2}\underline{k} + \underline{k}',t)$ and $C(\frac{1}{2}\underline{k} - \underline{k}',t)$, respectively,
but there is always a probability, which is again given by
$g(\underline{k},\underline{k}')$, that they will interfere and create a new fluctuation
with wavevector $\underline{k}$. This complicated process is described by
eq. (3.16). If $\tilde{C}(\underline{k},t) > 0$, which can be expected for overdamped
systems, as it is colloidal and polymeric solutions, the
"memory" is always a positive one, $M(\underline{k},t) > 0$. Therefore the
memory-term in eq. (2.8) will diminish the influence of the
meanfield term $D(\underline{k})k^2$, $\tau_m(\underline{k}) > \kappa_1(\underline{k})^{-1}$, and the long-time relaxation
of $C(\underline{k},t)$ will be slower than the short-time relaxation due to such
interference-processes.

Eqs. (3.16) and (3.17) are also suitable starting points for
calculations,both analytical as well as numerical ones, since the
only quantity which is needed to evaluate $M(\underline{k},t)$ is the static
structure factor $S(\underline{k})$. As long as such calculations are not
performed it would be interesting to see how from experiments the
memory-function can be extracted. The best method to do this would
be to Laplace-transform the experimentally measured $C(\underline{k},t)$,
calculate $\tilde{M}(\underline{k},z)$ by eq. (3.6) and Laplace-transform again $\tilde{M}(\underline{k},z)$.

A much easier way, although with limited accuracy, is to assume
a two-exponential form for $C(\underline{k},t)$ and to fit the experiments to it:

$$C(\underline{k},t) = C(\underline{k},0)(\alpha_1 \, e^{-|t|/\tau_1} + \alpha_2 \, e^{-|t|/\tau_2}) \, , \qquad (3.19)$$

which yields

$$\tilde{C}(\underline{k},z) = C(\underline{k},0)(\frac{\alpha_1}{z + \tau_1} + \frac{\alpha_2}{z + \tau_2}) \qquad (3.20)$$

and

$$\tilde{M}(\underline{k},z) = M(\underline{k},0) \, \frac{1}{z + \tau_I^{-1}} \qquad (3.21)$$

by using the memory equation.

Therefore $M(\underline{k},t))$ is an exponential,

$$M(\underline{k},t) = M(\underline{k},0) \, \exp\left[- |t|/\tau_I\right] \, , \qquad (3.22)$$

$$M(\underline{k},0) = -\kappa_2(\underline{k}) \cdot S(\underline{k}) = \alpha_1 \alpha_2 \cdot S(\underline{k}) \, \frac{(\tau_2-\tau_1)^2}{(\tau_1 \cdot \tau_2)^2} \qquad (3.23)$$

$$\tau_I = \frac{\tau_1 \cdot \tau_2}{\alpha_1 \tau_1 + \alpha_2 \tau_2} \qquad (3.24)$$

But it must be stressed that this result is produced by the ansatz (3.19) and only has to be considered as a pragmatic way to analyse the experimental data. The work of Dieterich and Peschel (2) as well as calculations of memory-functions in fluids (40) show that $M(\underline{k},t)$ is certainly not a simple exponential in the general case.

Grüner and Lehmann (32,33) have used such a two-exponential analysis to determine the relative deviation $(\kappa_1(\underline{k}) - \tau_m^{-1}(\underline{k})/\kappa_1(\underline{k})$, which characterizes the non-exponential behaviour, for systems of charged particles at different concentrations. They found that it is a universal function of k/k_m, where k_m is the wavevector where $S(\underline{k})$ has its maximum.

SELF DIFFUSION

A memory-equation for the Fourier transform of the self diffusion propagator of a particle i,

$$C_s(\underline{k},t) \equiv \, <\exp(-i \, \underline{k}(\underline{R}_i(t) - \underline{R}_i(0)))> \, , \tag{4.1}$$

can be derived (1,2), similarly as for the concentration auto-correlation function $C(\underline{k},t)$:

$$\frac{\partial}{\partial t} \, C_s(\underline{k},t) = -D_o k^2 \, C_s(\underline{k},t) + \int_0^t dt' \, M_s(\underline{k},t-t')C_s(\underline{k},t')$$
$$\text{for } t>0 \tag{4.2}$$

But contrary to the memory-function of the concentration fluctuations $M_s(\underline{k},t)$ does not vanish for small wavevectors if one neglects three-particle correlations. Therefore $M_s(\underline{k},t)$ contributes to the hydrodynamic limit of eq. (4.2) even in this approximation. Taking the Laplace transform of (4.2) gives

$$\tilde{C}_s(\underline{k},z) = \{z + D_o k^2 - \tilde{M}_s(\underline{k},z)\}^{-1}. \tag{4.3}$$

In the hydrodynamic limit one finds

$$\tilde{C}_s^h(\underline{k},z) = \lim_{\substack{z \to 0 \, k \to 0 \\ z/k^2 = const.}} \tilde{C}_s(\underline{k},z) = \{z + D_s k^2\}^{-1} \, , \tag{4.4}$$

so that

$$C_s^h(\underline{k},t) = e^{-D_s k^2 \cdot |t|} \tag{4.5}$$

with

$$D_s = D_o - \lim_{k \to 0} \tilde{M}_s(\underline{k},0)/k^2 \, . \tag{4.6}$$

Using the continuity equation we can again relate the velocity autocorrelation function to the self diffusion propagator,

$$\frac{1}{3} <\underline{V}_i(t)\ \underline{V}_i(0)> = \lim_{k \to 0} \frac{1}{k^2} \frac{\partial^2}{\partial t^2} C_s(\underline{k},t)$$

(4.7)

$$= 2\ D_o\ \delta(t) - \lim_{k \to 0} M_s(\underline{k},t)/k^2,$$

where the second equality follows if we use the inverse Laplace transform of (4.3). Eqs. (4.6) and (4.7) show that the Kubo-relation holds.

$$D_s = \frac{1}{3} \int_0^\infty dt\ <\underline{V}_i(t)\cdot\underline{V}_i(0)> \quad.$$

(4.8)

It is sometimes believed that the self diffusion coefficient can be related to the friction coefficient as $D_S = k_B T/f$, and therefore $D_c/D_S = \beta \cdot \frac{\partial \pi}{\partial c}\big|_T$. It is important to note that these relations are not valid. Although the self diffusion coefficient can be related to a friction coefficient, this is not the same friction coefficient as in the case of collective diffusion and it is not the friction cofficient measured in a sedimentation or electro-phoresis experiment. This can be seen in the following way.

The Fokker-Planck equation for the distribution function of coordinates and momenta of N Brownian particles   (41,42)

$$\frac{\partial}{\partial t}\ f_N(\{\underline{R}_\alpha,\underline{P}_\alpha\},t) = (\hat{O}(\{\underline{R}_\alpha,\underline{P}_\alpha\} - \sum_i \underline{F}_i \frac{\partial}{\partial \underline{P}_i})f_N(\{\underline{R}_\alpha,\underline{P}_\alpha\},t) \quad.(4.9)$$

$\underline{F}_i$ is a constant external force acting on particle i,

$$\hat{O}(\{\underline{R}_\alpha,\underline{P}_\alpha\}) = -\sum_{i=1}^N \left[\frac{P_i}{M}\frac{\partial}{\partial \underline{R}_i} - (\frac{\partial}{\partial \underline{R}_i} U_N(\{\underline{R}_\alpha\}))\frac{\partial}{\partial \underline{P}_i} \right]$$

(4.10)

$$+ \sum_{i,j=1}^N \frac{\partial}{\partial \underline{P}_i}\ \underline{\underline{\zeta}}_{ij}(\{\underline{R}_\alpha\})\ (\beta \frac{\partial}{\partial \underline{P}_j} + \frac{P_j}{M})\ ,$$

the $\underline{\underline{\zeta}}_{ij}(\{\underline{R}_\alpha\})$ are the elements of the inverse matrix of $\underline{\underline{D}}_{ij}(\{\underline{R}_\alpha\})$ times $k_B T$.

The linear response solution of (4.9) is

$$f_N^{(1)}(\{\underline{R}_\alpha,\underline{P}_\alpha\},t) = f_N^{(o)}(\{\underline{R}_\alpha,\underline{P}_\alpha\}) - \sum_i \int_0^t dt'\ e^{\hat{O}(\{\underline{R}_\alpha,\underline{P}_\alpha\})(t-t')}$$

$$\cdot \; \underline{F}_i \; \frac{\partial}{\partial \underline{P}_i} \; f_N^{(o)} (\{\underline{R}_\alpha, \underline{P}_\alpha\}) \quad , \tag{4.11}$$

where $f_N^{(o)}$ is the equilibrium solution,

$$f_N^{(o)} (\{\underline{R}_\alpha, \underline{P}_\alpha\}) = Z \cdot e^{-\frac{\beta}{2M} \sum_i P_i^2} \; e^{- \beta U_N (\{\underline{R}_\alpha\})} \quad , \tag{4.12}$$

and Z the normalization factor.

The mean velocity of a particle α becomes

$$\langle \underline{V}_\alpha (t) \rangle = \int d\{\underline{R}_\alpha, \underline{P}_\alpha\} \; \underline{V}_\alpha \cdot f_N^{(1)} (\{\underline{R}_\alpha, \underline{P}_\alpha\}, t)$$

$$= \beta \sum_{i=1}^N \int_o^t dt' \langle \underline{V}_\alpha \; e^{\hat{O}(\{\underline{R}_\alpha, \underline{P}_\alpha\})(t-t')} \; \underline{V}_i \rangle \cdot \underline{F}_i$$

$$= \beta \sum_{i=1}^N \int_o^t dt' \; \langle \underline{V}_\alpha (t-t') : \underline{V}_i (0) \rangle \cdot \underline{F}_i \quad . \tag{4.13}$$

For long times the stationary mean velocity is

$$\langle \underline{V}_\alpha (\infty) \rangle = + \beta \sum_{i=1}^N \int_o^\infty dt' \; \langle \underline{V}_\alpha (t') : \underline{V}_i (0) \rangle \cdot \underline{F}_i \quad . \tag{4.14}$$

We now consider two different situations:

I) $\underline{F}_i$ acts on all particles in the same way. This is the case in a sedimentation or electrophoresis experiment with identical particles. The inverse friction or drag coefficient is

$$f^{-1} = \frac{1}{3} \beta \frac{1}{N} \sum_{i,j=1}^N \int_o^\infty dt \; \langle \underline{V}_i (t) \cdot \underline{V}_j (0) \rangle \quad . \tag{4.15}$$

This is the friction coefficient considered in the section on collective diffusion.

II) $\underline{F}_i$ acts only on particls α. Then we can define a "self-friction" coefficient as

$$f_s^{-1} = \frac{1}{3} \beta \int_o^\infty dt \; \langle \underline{V}_\alpha (t) \cdot \underline{V}_\alpha (0) \rangle \quad . \tag{4.16}$$

We see by comparing this relation with eq. (4.8) that it is

such a "self-friction" coefficient, which is related to the
self diffusion coefficient. It gives the friction on a
particle, when only this particle is moved by an external
force. Since it is difficult to imagine how such an
experiment could be performed this "self-friction" coefficient
seems to be an artificial quantity.

We see from this discussion that it is not possible to relate
the sedimentation coefficient to the self diffusion coefficient
in a simple manner.

For the memory function of the self diffusion process we can also
give a mode-mode coupling approximation (39), neglecting again
hydrodynamic interactions this takes the form*

$$M_s(\underline{k},t) = \frac{c}{(2\pi)^3} \int d^3k' \{D_o\underline{k}\cdot\underline{k}' \; h_D(\underline{k}')\}^2$$

$$\cdot \; C_s(\underline{k}-\underline{k}',t) \cdot C(\underline{k},t) \quad . \tag{4.17}$$

This describes the interaction between a concentration fluctuation
and the self diffusion propagator, the strength of the interaction
is again proportional to the direct correlation function h_D, eq.
(3.18). If we interpret the concentration fluctuation as a
cloud of particles, M_s gives the probability that a single
particle, whose motion is described by C_s, enters the cloud at
t = 0, moves with the cloud until time t and separates from it.
We would expect that the diffusion of the particle is slowed down
by this process and it is obvious that eq. (4.17) is nothing else
than the mathematical formulation of Pusey's cage model (29).

The memory-function in the approximation (4.17) has been
evaluated for a weakly coupled system of charged particles (39).
If one assumes that the charge of a Brownian particle is much
larger than the charge of the counterions one finds

$$D_s = D_o \{1 - 0.346 \cdot (\frac{\beta}{\varepsilon})^{3/2} \; Q^3 \; c^{1/2}\} \quad , \tag{4.18}$$

where ε is the solvent dielectric constant.

The same result has been found by Onsager (43) and Harris (44)
by quite different methods.

*But note that the exact expression for the memory function
contains the hydrodynamic interaction (1) and therefore D_s is
influenced by it.

SUMMARY

    It was the intention of this paper to show how, in concentrated
solutions of interacting particles, diffusion coefficients, friction
coefficients and other related quantities can be expressed by the
static distribution functions and the hydrodynamic interaction in
a general way. We have seen that the differences between various
theoretical results are not so much due to their different starting
points but to the fact that they treat different systems and
use different approximations for the hydrodynamic interactions.

    The question as to the correct form of the hydrodynamic
interaction is a difficult one, since hydrodynamic interaction in
a concentrated system is a complicated many-body process. The
simplest and most frequently used approximation is the Oseen
tensor (2.22). The Oseen tensor should be valid for two isolated
particles whose distance is much larger than their radii. This
approximation is therefore unsatisfactory when the expectation value
$\langle \underline{D}_{ij}(r) \rangle$ is dominated by the short-distance behaviour of the
hydrodynamic interaction, as it is the case at low concentrations
with short range correlations, and for highly concentrated
systems where the two-particle approximation is no longer valid.
For the first kind of systems hydrodynamic interaction functions
which are also valid at short distances have been given by, e.g.
Batchelor (12) and Felderhof (13). For the second kind of systems
a screened version of the Oseen tensor has been proposed by Freed
and Edwards (45), De Gennes (23) and Adelman (46). Many experiments
on colloidal and polymeric solutions show a pronounced non-
exponential time behaviour. Here the correlation function decays
for long times much slower than for short times. It has been
shown how this can be related to a memory function. The latter
is interpreted as a quantity which gives the probability for the
coupling of two concentration fluctuations or the coupling of a
concentration fluctuation to the self diffusion propagator in the
case of the self diffusion memory function, respectively. The
time-integrated memory-function has been extracted from experiments
for systems of charged particles (32,33). For five systems with
concentrations varying by a factor of 10, it has been found to be
a function of k/k_m, and k_m^{-1} is proportional to the mean distance
between two nearest neighbours. It is well known that in such
systems of charged particles there exists a transition to a
crystalline structure (14) and it would be especially interesting
to see whether the time integral of the memory-function is still
such a universal function of k/k_m near the transition point or
whether it influences the dynamics of the phase transition.

ACKNOWLEDGEMENT

I am grateful to Prof. R. Klein for many valuable discussions and for critical reading of the manuscript.

REFERENCES

1. B. J. Ackerson, J. Chem. Phys., 69:684 (1978).
2. W. Dieterich and J. Peschel, Physica 95A:208 (1978).
3. W. Hess and R. Klein, Physica 94A:71 (1978).
4. W. Hess and R. Klein, to be published.
5. A. R. Altenberger, Chem. Phys., 15:269 (1976).
6. J. G. Kirkwood et al., J. Chem. Phys., 33:1505 (1960).
7. G. Phillies, J. Chem. Phys., 60:976 (1974).
8. B. U. Felderhof, J. Phys., A11:929 (1978).
9. C. W. Pyun and M. Fixman, J. Chem. Phys., 41:937 (1964).
10. A. R. Altenberger and J. M Deutch, J. Chem. Phys., 59:894 (1978).
11. W. Hess and R. Klein, Physica 85A:509 (1976).
12. G. R. Batchelor, J. Fluid Mech., 52:245 (1972).
13. B. U. Felderhof, Physica 89A:373 (1977).
14. R. Williams and R. S. Crandall, Phys. Lett., 48A:225 (1974).
15. G. R. Batchelor, J. Fluid Mech., 74:1 (1976).
16. A. R. Altenberger, J. Chem. Phys., 70:1994 (1979).
17. M. J. Stephen, J. Chem Phys., 55:3878 (1971).
18. P. Doherty and G. B. Benedek, J. Chem. Phys., 61:5426 (1974).
19. R. Schor and E. N. Serrallach, J. Chem. Phys., 70:3012 (1979).
20. M. Adam, M. Delsanti and G. Jannink, J. Phys. Lett., 37:L53 (1976).
21. M. Adam and M. Delsanti, Macromolecules 10:1229 (1977).
22. P. G. De Gennes, Macromolecules, 9:587 (1976).
23. P. G. De Gennes, Macromolecules, 9:594 (1976).
24. Z. Akcasu and H. Gurol, J. Polym. Sci., Polym. Phys. Ed., 14:1 (1976).
25. M. Benmouna and Z. Akcasu, Macromolecules, 11:1187 (1978).
26. Z. Akcasu and M. Benmouna, Macromolecules, 11:1193 (1978).
27. H. Yamakawa, Modern Theory of Polymer Solutions, Harper and Row, New York, (1971).
28. Judith C. Brown, P. N. Pusey, J. W. Goodwin and R. H. Ottewill, J. Phys., A8:664 (1975).
29. P. N. Pusey, J. Phys., A11:119 (1978).
30. P. S. Dalberg, A. Boe, K. A. Strand and T. Sikkeland, J. Chem. Phys., 69:5473 (1978).
31. P. N. Pusey in these proceedings.
32. F. Grüner and W. Lehmann, to be published in J. Phys. A
33. F. Grüner and W. Lehmann in these proceedings.
34. G. Jones and D. Caroline, Chem. Phys., 40:153 (1979).

35. D. Richter, J. B. Hayter, F. Mezei and B. Ewen, Phys. Rev. Lett., 41:1484 (1978).

36. D. Caroline in these proceedings.

37. P. N. Pusey, J. Phys., A8:1433 (1975).

38. B. J. Ackerson, J. Chem. Phys., 64:242 (1976).

39. W. Hess and R. Klein, to be published.

40. Y. Pomeau and P. Résibois, Physics Reports, 19:63 (1975).

41. J. M. Deutch and J. Oppenheim, J. Chem. Phys., 54:3547 (1971).

42. T. J. Murphy and J. L. Aguirre, J.Chem Phys., 57:2098 (1972).

43. L. Onsager, Ann. N. Y. Acad. Sci., 46:241 (1945).

44. S. Harris, Mol. Phys., 26:953 (1973).

45. K. F. Freed and S. F. Edwards, J. Chem. Phys., 61:3626 (1974).

46. S. A. Adelman, J. Chem. Phys., 68:49 (1978).

ON THE LONG TIME DIFFUSION OF INTERACTING BROWNIAN PARTICLES

F. Grüner and W. Lehmann

Universität Konstanz
Fakultät für Physik
Konstanz, Federal Republic of Germany

ABSTRACT

The short and long-time diffusion coefficient in suspensions
of charged polystryrene spheres was determined by photon correlation
spectroscopy over a wide range of concentrations and k-vectors.
The results indicate, that the long time diffusion of interacting
Brownian particles may be described by the static structure factor
which gives the concentration dependence and a reduced memory
function $M'(k,\omega=0)$ which is independent of the concentration and
describes the dynamical aspects of the repulsive interaction
potential.

INTRODUCTION

The method of quasielastic light scattering has become a very
widely used tool to determine the diffusion coefficient and hence
the size of macromolecules in solution. This is however restricted
to sufficiently diluted systems. In more concentrated systems two
effects occur and influence the spectrum of the scattered light.
The first one is due to the occurance of multiple scattering
events which bears different and in most cases unwanted information.
The second effect is the onset of interaction between the
macromolecules giving interesting information on the diffusion
process in the presence of interaction which may help to understand
the diffusion processes in biological and other high concentrated
systems. Two model systems of interacting particles have been
studied, which represent the two extremes of possible repulsive
interaction potentials. One is nearly a hard core system (1) and
the other, which will be studied here to, is a system of charged
polystyrene spheres (2, 3, 4, 5). The latter interacts due to the

long range Coulomb potential and allows moderate particle
concentrations of about 10^{-3} per volume to study the influence of
interaction. The surface charge on the polystyrene spheres is
produced by the dissociation of about 1000 protons into the water
and gives rise to liquid or even crystalline ordering. The
corresponding static structure factors S(k) could be determined
by a static light scattering experiment (2, 6) (Fig. 1).

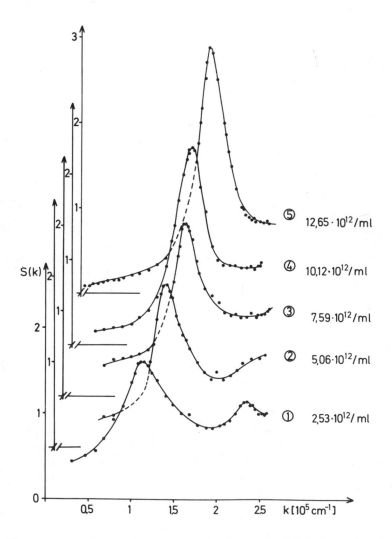

Fig. 1. The measured static structure factors S(k) for the five
         samples used. The concentrations are given in particles
         per ml.

It was shown experimentally (2) and theoretically (3) that in the short time limit the diffusion coefficient is given by $D_{eff} = D_o/S(k)$ where D_o is the free particle diffusion coefficient. Recently Pusey (4) and Dalberg et al (5) observed a long time tail in the autocorrelations-function of the scattered light. We have studied the concentration and k-vector dependence of this long time tail in more detail and will discuss the results in the frame of a memory function formalism.

The outline of our paper is as follows. In the next chapter we give the experimental details. Then we give a detailed discussion of the method used to correct the static and dynamic experiments for multiple scattering. In the last chapter we present the results and evaluate the memory function in the limit $\omega \to 0$.

Experimental

a)   Sample preparation

We used polystyrene spheres of radius 0.045µ, supplied by DOW. They were suspended in filtered (0.45µ Millipore) bidestilled water with mixed bed ion exchange resin (MB 2 DOW-EX) added. By this preparation most of the counterions besides the protons from the particles are removed. After several days the suspensions show static structure factors different from one, indicating the onset of interaction. Five different concentrations from 2.53 10^{12} to 12.65 10^{12} particles/ml ($1^o/oo$ to $5^o/oo$per volume) were prepared. The static structure factors measured are given in Fig. 1.

From the measured pH-value a surface charge of $1100 \pm 150e^-$ can be calculated assuming that all protons in the suspensions come from the latex. Applying the ordinary Debye Hückel theory a screening length of the order of several Angstroms is obtained which is in contradiction to the mean particle distance of several thousand Angstrom estimated from the peakposition of the static structure factors. Such a characteristic length of a screened Coulomb potential can be obtained if one assumes an effective charge much smaller than the actual charge. This might be due to a strongly coupled proton cloud surrounding the Latex which reduces the surface charge. This is in agreement with a recent paper (7) where an effective charge of about 100 e^- is calculated.

We also checked the concentration dependence of the peak of S(k) (Fig. 2) which varies like

$$k_{max} \propto \sqrt[3]{n} \qquad\qquad (1)$$

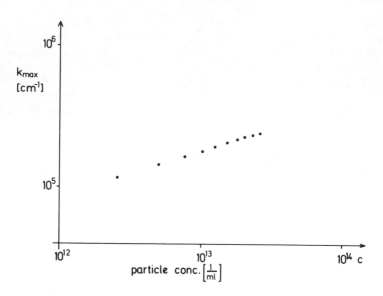

Fig. 2.   Log-log plot of the peak position of S(k) versus the
          particle concentration.   Wavelength used 632.8 nm.

as it is expected; n is the particle concentration.   This is even
true for the crystalline like structure factors which the five
highest concentrations in figure 2 show.

(b)   Apparatus

     The sample cells used are cylindric quarz tubes with an
internal diameter of 6 mm.   They are surrounded by a bath of
glycerin, temperature stabilized to 25 ± 0.5°C, which also acts
as an index-matching fluid.   To cover a wide range of k-values
we used either the 632.8 nm He-Ne-laserline or the 457,9 nm
Ar-laserline.   The scattered light was collected on a RCA-
photomultiplier (type C 31034).   For the static measurements we
used an optical arrangement of two intersecting parallel beams and
conventional photocounting equipment.   For the dynamic measurements
the scattering volume was confined by the waist of the focussed
laser with a diameter of about 200 μ and a 1 : 1 projection of a
30 μ diameter pinhole in front of the photocathode by a lens with
20 cm focal length.   To achieve neat polarisation geometry Glan-
Tompson prisms with an extinction ratio of better than 10^{-6} were
used both as polarizer and as analyser. We used a minicomputer
based 4000 channel correlator (8) which determines the full intensity
correlation function.

   Multiple scattering and data analysis. All data analysis
was done by fitting several exponentials using a multiexponential
fit program developed by Provencher (9). There are no input
parameters to this fit procedure except the experimental data
and the maximum number of exponentials. We always get a good fit
within the experimental error. This assures, that the fitted
function represents the measured function reasonably well although
theoretical calculations may give another functional form of the
correlations-function. The model functions thus obtained will
facilitate further calculations since one deals with a limited
number of parameters.

   The crucial point in the data analysis is the correction for
multiple scattering since these effects disturb the measurements
heavily essentially at high concentrations and low scattering
angles. A method for the correction was described in an earlier
peper (10) and will be briefly outlined here for completeness.
The basic idea is to exploit all scattering geometries available
(see fig. 3 for notation).

   In VV geometry the normal single scattered light together
with the parallel polarized component of the multiple scattered
light is measure In VH geometry the depolarized component of
the multiple scattered light shows up alone. In HH geometry at
90° scattering angle one is able to measure, provided small
spherical particles, the parallel polarized component of the
multiple scattering alone due to the dipolar nature of the
scattering process. Measuring the depolarized VH (HV) scattering

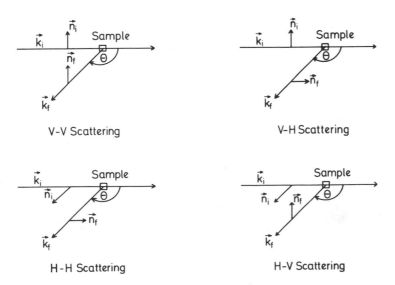

V-V Scattering   V-H Scattering

H-H Scattering   H-V Scattering

Fig. 3. Definition of scattering geometries.

together with the HH scattering at $\Theta = 90^{o}$ one gets a reasonable amount of information to do good corrections for multiple scattering. The procedure is as follows:

a)   Static Measurements

A detailed knowledge of the static structure factor $S(k)$ is necessary to crosscheck the short time diffusion coefficient. If we denote the single scattering by 1 and the multiple scattering by n, the following equations hold

$$I_{VV}^{tot}(\Theta) \;=\; I_{VV}^{1}(\Theta) \;+\; I_{VV}^{n}(\Theta)$$

$$I_{VH}^{tot}(\Theta) \;=\; I_{HV}^{tot}(\Theta) \;=\; I_{VH}^{n}(\Theta) \tag{2}$$

$$I_{HH}^{tot}(90^{o}) \;=\; I_{HH}^{n}(90^{o})$$

If we assume that there is no difference in the angular dependence of polarized and depolarized multiple scattered component, we get the following equation:

$$I_{VV}^{n}(\Theta) \;=\; I_{HH}^{n}(\Theta) \;=\; \frac{1}{R^{n}}\, I_{VH}^{n}(\Theta) \tag{3}$$

where the constant depolarization ratio R^{n} may be determined from the intensities at $\Theta = 90^{o}$

$$R^{n} \;=\; \frac{I_{HV}^{n}(90^{o})}{I_{HH}^{n}(90^{o})} \tag{4}$$

These considerations yield the following expression for the static structure factor

$$S(k) \;=\; I_{VV}^{1}(\Theta) \;=\; I_{VV}^{tot}(\Theta) - \frac{1}{R^{n}}\, I_{VH}^{n}(\Theta) \tag{5}$$

All data given in figure 1 are corrected in this fashion. It should be noted, that this correction becomes very important in forward scattering. For example the multiple scattered intensity at $\Theta = 30^{o}$ of sample 5 was 58% of the total intensity.

b)   Dynamical Measurements

It has been shown (10, 11, 12, 13), that the multiple scattering tends to produce short decay times in the autocorrelation

function. It is therefore possible to identify the multiple
scattered components in a multi-expenential fit. From a comparison
of the parallel polarized and depolarized multiple scattering
correlation functions (HH and HV geometries at $\theta = 90^{\circ}$) (10) it is
known, that the decay times of the parallel component are only
about 10% larger than in the depolarized case, so that the decay
times determined from the VH geometry are sufficient to make an
unambiguous choice of the multiple scattered components in the VV
correlation function. We will demonstrate the procedure by an
example. We choose the measurement from sample 5 at $\theta = 30^{\circ}$. As
mentioned above, the multiple scattered light is more than 50% in
this case. The fit results for the polarized and depolarized
correlations function are given in table 1.

TABLE I

Results of the multiexponential fit for the polarized and depolarized
intensity-correlation function g^2_{VV} and g^2_{VH} respectively of sample
5 at 30° scattering angle.

The amplitudes are given in units of the flat background.

Intensity - Correlations Function

Polarized g^2_{VV}		Depolarized g^2_{VH}	
Amplitude	Decay time μsec	Amplitude	Decay time μsec
0.016	1,269 ± 0,180		
0.085	308 ± 35	0.061	305 ± 45 +
0.141	105 ± 17	0.217	99 ± 10
0.085	37 ± 6	0.127	30 ± 5

Note that the results are from one run with a sample time (time
delay per channel) of 1.28 μ sec resulting in a maximum delay time
of 5.2 msec. There are four exponentials in VV-scattering and
three exponentials in HV-scattering with a very nice agreement of
the shorter decay times. Care has to be taken however with the 308
μ sec component since the static structure factor is very small
and a short time constant in this region is expected. Looking
at the intensity ratios of 1 : 1.8 : 0.5 in the VH- and

1 : 1.7 : 1 in VV polarisation for these components it is clear that
the 300 µs component has parts from both, multiple and single
scattering in a ratio of about 1 : 1. So the correction for multiple
scattering is done by subtracting of the two shortest components as
a whole and of 50% of the 308 µsec component. The resulting
correlation function should be a good representation of the single
scattered light. We would like to stress that this is an extreme
example with respect to the high multiple scattered intensity and
the very low static structure factor. The good agreement of the
first cumulant of the field correlation function and the static
measurements (fig.7) however shows that our procedure gives correct
results. Neglecting the multiple scattering effects would give an
enormous disagreement between S(k) and $\dfrac{D_{eff}}{D_o}$ $^{-1}$.

c) Analysis of the field-correlation function

    The corrected intensity-correlation function is normalized with
a background corresponding to the single scattered part of the
total intensity. Assuming the applicability of the Siegert
relation which will be discussed in the next chapter we calculate
the field-correlation function. We show ten measurements in
semilogarithmic plots together with the fits obtained. Figure 4
shows the correlation function at an angle of $\theta = 30^o$ for all
samples. This angle corresponds always to a k-vector well below
the peak. The fitparameters obtained are listed in table 2. The
same is given for scattering angles at the peak of S(k) in
figure 5 and table 3 respectively.

    It is clearly seen that the long tail has an exponential decay
and that the data are well fitted by two exponentials over the
total timescales. The deviations in the intermediate time region
between the two exponentials observed by Pusey (4) might be due to
the fact that these data were not corrected for multiple scattering.

    Results and discussion. In the presence of interactions between
macromolecules in solution, the collective diffusion coefficient is
affected considerably, both in magnitude and its time dependence.
There are two timescales involved which are defined by two collision
times, the typical time describing collisions between the Brownian
particle and the solvent molecules τ_c and the collision time τ_{int}
between the Brownian particles themselves. Whereas τ_c is estimated
to be of the order of 10^{-10} sec which is not covered by photon
correlation technique, τ_{int} is defined by the motion of the particles
and leads to a random interacting force varying on the same timescale as the diffusion itself. At times $\tau_c << t << \tau_{int}$ the interaction
field from the particles may be regarded as static and approximated
by its mean field value. On this timescale the diffusion coefficient
is given by the well established (2, 3) relation.

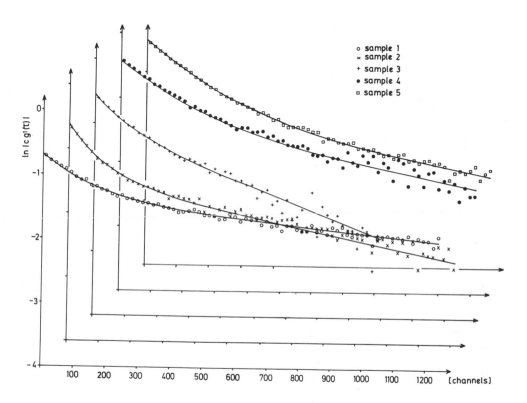

Fig. 4.   Some field correlation functions at $\Theta = 30^{\circ}$ with
$\lambda = 632.8$ nm in semilog-plot together with the fitted
functions. The timescale is given in channels of the
correlator. To convert to μsec multiply with the
sampletime given in the table below. Note that only a
small fraction of data points are plotted for clarity.

TABLE II

The Parameters of The Fit Curves
in Figure 4

	Amplitude A_1	Decayconst Γ_1 μsec^{-1}	Amplitude A_2	Decayconst Γ_2 μsec^{-1}
Sample 1 $\Theta = 30^\circ$ Sample time 2.56 μsec	.243	$1.17 \cdot 10^{-4}$	.242	$1.48 \cdot 10^{-3}$
Sample 2 $\Theta = 30^\circ$ Sample time 2.56 μsec	.233	$2.59 \cdot 10^{-4}$	.308	$2.57 \cdot 10^{-3}$
Sample 3 $\Theta = 30^\circ$ Sample time 2.56 μsec	.343	$3.47 \cdot 10^{-4}$	.230	$1.96 \cdot 10^{-3}$
Sample 4 $\Theta = 30^\circ$ Sample time 1.28 μsec	.258	$3.56 \cdot 10^{-4}$	.393	$2.28 \cdot 10^{-3}$
Sample 5 $\Theta = 30^\circ$ Sample time 1.28 μsec	.209	$3.57 \cdot 10^{-4}$	.409	$2.18 \cdot 10^{-3}$

$$D_{eff} = \frac{D_o}{S(k)} \tag{6}$$

At longer times the random walk of the particles leads to fluctuations of the interaction force on the timescale τ_{int}. This is most conveniently described in terms of a memory function in the generalized Langevin equation and leads to the following equation of motion for the time correlation function:

$$\frac{d}{d\tau} g(\tau) = -\Omega(k) g(\tau) - \int_0^\tau dt\ M(t)\ g\ (\tau-t) \tag{7}$$

where

$$\Omega(k) = \frac{D_o k^2}{S(k)} = D_{eff} k^2$$

The memory function introduces a non Markov process into the system which might affect the applicability of the Siegert relation. This contribution vanishes at small times where the integral is zero. If we assume that $M(t)$ decays faster than $g(\tau)$ at large times the Markov approximation is valid and hence the Siegert relation gives correct answers at least in the short and long term limits.

Following Dieterich and Peschel (14) the dynamic structure factor is given by (see also Hess (15))

$$S(k,\omega) = \frac{1}{\pi}\ Re\ \frac{S(k)}{-i\omega + \Omega(k) + M(k,\ -i\omega)} = \frac{1}{\pi}\ \frac{S(k)\{\Omega(k) + M(k,-i\omega)\}}{\omega^2 + \{\Omega(k) + M(k,-i\omega)^2\}} \tag{8}$$

where $M(k,z)$ is the Laplace transform of the memory function defined in equ.2. The collective diffusion coefficient in the hydrodynamic limit is defined (14)

$$D = \lim_{k\to o}\ \lim_{\omega\to o}\ \left[\frac{D_o}{S(k)} + \frac{M(k,\ -i\omega\ )}{k^2}\right] \tag{9}$$

from which we may define a k-dependent long time diffusion coefficient

$$D_L = \lim_{\omega\to o}\ \left[\frac{D_o}{S(k)} + \frac{M(k,\ -i\omega)}{k^2}\right] \tag{10}$$

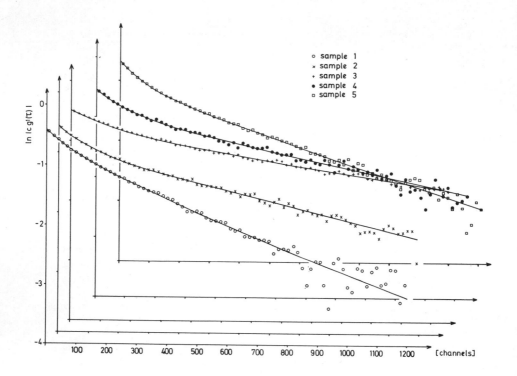

Fig. 5.   Same as figure 4 for the maximum of S(k). Data are given
          in the table below.

which is not the decay constant of the long time tail of the
correlation function, in contrast to previous interpretations
(4,5). It is straightforward to calculate D_L from the model
functions used to fit the experimental field-correlation function
$g^1(\tau)$

$$g^1(\tau) = \sum_i \alpha_i \, e^{-\Gamma_i \tau} \quad \text{with} \quad \sum_i \alpha_i = 1 \tag{11}$$

TABLE III

The Parameters of the Fit Curves
in Figure 5

	Amplitude A_1	Decayconst Γ_1 μsec^{-1}	Amplitude A_2	Decayconst Γ_2 μsec^{-1}
Sample 1 $\Theta_{Max}=50°$ Sample time 5.12 μsec	0.505	$2.237.10^{-4}$	0.1434	$1.25 \cdot 10^{-3}$
Sample 2 $\Theta_{Max}=65°$ Sample time 2.56 μsec	0.390	$2.373.10^{-4}$	0.181	$2.19 . 10^{-3}$
Sample 3 $\Theta_{Max}=77°$ Sample time 1.28 μsec	0.436	$2.902.10^{-4}$	0.176	$2.45 \cdot 10^{-3}$
Sample 4 $\Theta_{Max}=88°$ Sample time 2.56 μsec	0.386	$2.07 .10^{-4}$	0.1877	$1.85 \cdot 10^{-3}$
Sample 5 $\Theta_{Max}=95°$ Sample time 2.56 μsec	0.339	$3.07.10^{-4}$	0.13	$2.36 \cdot 10^{-3}$

The Fouriertransform yields for the dynamic structure factor a sum of Lorentzians:

$$S(k,\omega) \;=\; \frac{1}{\pi}\, S(k,0)\, \Sigma \frac{\alpha_i \Gamma_i}{i\omega^2 + \Gamma_i^2} \qquad\qquad (12)$$

Combining this with equation (8) and (10) gives:

$$D_L k^2 \;=\; \frac{1}{\displaystyle\sum_i \frac{\alpha i}{\Gamma_i}} \qquad\qquad (13)$$

The short time diffusion coefficient is given by the initial slope of $g^1(\tau)$ to be

$$D_S k^2 \;=\; \sum_i \alpha_i \Gamma_i \qquad\qquad (14)$$

In figure 6 the k-dependence of these quantities are given for sample 1 together with the longest component $\tau_L k^2$ of the fit. Note that D_L^{-1} has a k dependence different of $S(k)$ whereas D_S^{-1} follows very satisfactory the static structure factor. If one looks to smaller k-values even the long time tail coincides with the short time behaviour, as can be seen from figure 7 where the same quantities are shown for sample 5.

From the measured short and long time diffusion coefficients the shape of the memory function may be extracted as it has been done in a previous paper (16). In the limit of $\omega \rightarrow o$ the memory function is given according equ. (10) by

$$M(k,\,\omega{=}o) \;=\; (D_L - D_S)k^2 \qquad\qquad (15)$$

If we define a reduced memory function $M^1(k,\,\omega{=}o)$ by

$$M^1(k,\,\omega{=}o) \;=\; -\,\frac{M(k,\,\omega{=}o)}{\dfrac{D_o k^2}{S(k)}} \;=\; \frac{D_S - D_L}{D_S} \qquad\qquad (15a)$$

all data from the five samples fit the same universal function if we plot it as function of the reduced wavevector k/k_{max}, where k_{max} is the peak position of the static structure factor. This surprising result leads to the conclusion, that the shape of M^1 depends only on the interaction potential whereas the concentration dependence enters only in k_{max}.

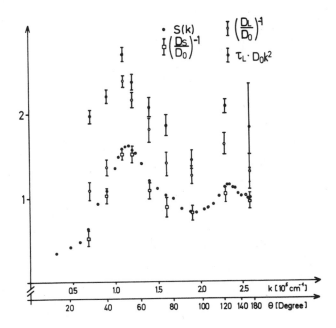

Fig. 6.   The measured static structure factor, the reciprocal short time diffusion coefficient, the reciprocal long time diffusion coefficient and the longest component of the fit τ_L . k^2 in units of D_o for sample 1.

Also included in fig. 8 are the data form Ref. 4 obtained with smaller particles ($r = 0.025\mu$) analysed according equ. (15a) once without correction for multiple scattering once with a crude correction where we have used the estimation of the influence of multiple scattering on the first cumulant given by Ref. 4. The satisfactory agreement of the corrected values with our results demonstrates again the importance of the multiple scattering correction. We conclude further, that the particle size has a minor influence on the reduced memory function, or in other words, the assumption of pointlike particles is justified.

From the log-log plot of $M^1(k, \omega=o)$ given in fig. 9 it is seen that M^1 dies off from its maximum with an approximate power low of

$$\frac{k}{k_{max}}\ ^{-1}$$

whereas for small k values no simple power law may be extracted over a wider range of k values. A power of 1 gives a reasonable approximation of M^1 up to a k-value of about $0.3\ k_{max}$. (Compare fig. 8 with the point at k = 0.06 not included in fig.9).

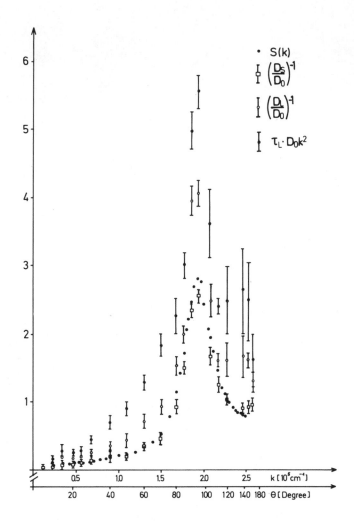

Fig. 7.   Same as figure 6 for sample 5.

The pronounced peak at $k = k_{max}$ seems to be reasonable since a
length scale of the order of the mean particle distance is the
typical length of the density fluctuations and in this region the
strongest influence is expected. At larger k-values, that is at
shorter characteristic length, the simple picture of a particle
fluctuating in a "cage" of it neighbours would imply that the interaction effects get stronger and the memory function increases
approaching k_{max}. At k-values before and around k_{max} this picture no
longer holds since the corresponding length of the fluctuation
involves several particles. At very long characteristic length, that
is k→0, a mean over a large number of particles is measured.

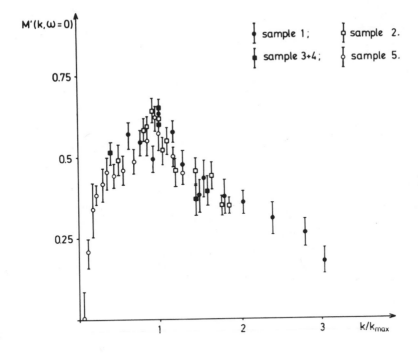

Fig. 8. The reduced memory function $M^l(k, \omega=0)$.

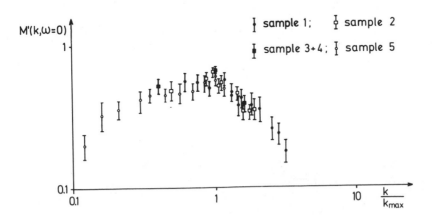

Fig. 9. Log-log plot of the memory function $M^l(k, \omega=0)$.

The diffusion is then conveniently described by the mean field value $D_0/S(k)$ and a memory function vanishing at $K \to 0$ is expected which is indeed observed. Calculations (14) for a one dimensional hard core system at low concentrations predict a power law of one in the limit of small k-values.

CONCLUSIONS

We have shown that in the long time limit the dynamics of interacting Brownian particles may be described by the static structure factor S(k) which gives the concentration dependence and a reduced memory function $M^l(k, \omega=0)$ which is independent of concentration. Hence $M^l(k, \omega=0)$ depends only on the interaction potential and describes the dynamical aspects of the repulsive interaction. The evaluation of this function at finite frequencies will be the subject of a further publication. We also showed that this type of measurement is extremely sensitive to distortions due to higher scattering effects and a lot of care has to be taken to analyse the experimental data, which requires the knowledge of all the informations of the scattered light.

ACKNOWLEDGEMENT

The authors are indepted to Dr. Heß and Prof. Klein for many useful discussions and suggestions. We thank Prof. Weber for his support. The numerical calculations were performed on the TR 440 computer at the "Rechenzentrum der Universität Konstanz".

References

1. H. N. Fijnaut, C. Pathmamanoharan, E. A. Nieuwenhuis, A. Vrij, Chem.Phys. Lett., 59:351 (1978).
2. J. C. Brown, P. N. Pusey, J. W. Goodwin, R. H. Ottewill, J.Phys., A8:664 (1975).
3. P. N. Pusey, J.Phys., A8:1433 (1975).
4. P. N. Pusey, J.Phys., A11:119 (1978).
5. P. S. Dahlberg, A. Bøe, K. A. Strand, T. Sikkeland, J.Chem. Phys., 69:5473 (1978).
6. R. Williams, R. S. Crandall, Phys.Lett., 48A:225 (1974).
7. Hastings, J.Chem.Phys., 68:678 (1978).
8. W. Lehmann, to be published.
9. Provencher, J.Chem.Phys., 64:2772 (1976).
10. F. Grüner, W. Lehmann, submitted to J.Phys.A.
11. C. M. Sorensen, R. C. Mockler, W. J. O'Sullivan, Phys.Rev., A14:1520 (1976).
12. C. M. Sorensen, R. C. Mockler, W. J. O'Sullivan, Phys.Rev., A17:2030 (1978).
13. Böheim, W. Hess, R. Klein, Z.Phys., B32:237 (1979).
14. W. Dieterich, J. Peschel, Physica 94A:208 (1975).

15. W. Hess, Contribution in this procedings.
16. F. Grüner, W. Lehmann, J.Phys., A12 Nr.11 (1979).

THE APPLICATION OF BROWNIAN DYNAMICS TO PHOTON CORRELATION

SPECTROSCOPY

K. J. Gaylor, I. K. Snook, W. J. van Megen
and *R. O. Watts

Department of Applied Physics
Royal Melbourne Institute of Technology
Melbourne, Australia

INTRODUCTION

Some of the most interesting and useful properties of colloidal dispersions are manifested in their dynamical behaviour. For example, the flow of paints, inks and mineral slurries are of great industrial importance as are the diffusion and flow properties of biological systems in the life sciences.

Recently Photon Correlation Spectroscopy (PCS) has been used to measure the temporal autocorrelation function(ACF) of the intensity of light scattered by colloids (1), from which many of the dynamical properties may be obtained. In principle one can find, amongst other things, diffusion constants and static and dynamic structure factors from the ACF. For example, for a system of identical non-interacting spherical particles, the ACF is a single exponentially decaying function of time (1). In this case the results may be unambiguously interpreted in terms of particle diffusion constants and hydrodynamic radii. However, for more physically interesting systems the complex nature of the particle dynamics prevents the interpretation of the ACF in terms of simple models.

Computer simulation techniques provide essentially exact results for a given interparticle potential (2) and is thus ideally

*Department of Chemistry, Australian National University, Canberra, Australia.

suited to provide insight into the fundamental particle motions contributing to experimental observations, and thus aid in the interpretation of PCS data. As an illustration of the application of machine simulation techniques to colloidal dynamics we use the numerical Brownian Dynamics method of Ermak (3) to simulate the Brownian motion of spherical colloidal particles interacting via a screened Coulomb potential.

BROWNIAN DYNAMICS METHOD

The method of Brownian Dynamics (BD) was first introduced by Ermak and co-workers (3) and is a mean of simulating the Brownian motion of systems of particles by sampling from the probability distribution function given by the solutions of the Smoluchowski equation. It may be shown that the displacement of a Brownian particle over a time step Δt, subject to the assumptions

(i)   that co-operative hydrodynamic effects may be ignored,

(ii)   that the force on a particle is essentially constant over Δt, and

(iii)   that all momentum variables have relaxed away, is

$$\underline{r}_i(t + \Delta t) = \underline{r}_i(t) + \underline{R}_i(\Delta t) + \frac{D_o}{kT} \underline{F}_i(t)\Delta t, \tag{1}$$

where $r_i(t)$ is the position of the ith particle at time t. The term $\underline{R}_i(\Delta t)$ is the random displacement of the ith particle due to the interaction of the colloidal particle with all the solvent molecules and is found from the relationship

$$<\underline{R}_i^2(\Delta t)> = 6D_o\Delta t \tag{2}$$

and $\underline{F}_i$ is the force felt by the ith particle due to all the other colloidal particles.

Over the time step relevant to interacting colloidal particles, assumptions (ii) and (iii) will generally be valid and assumption (i) may be eliminated albeit at the expense of adding complicated extra terms to equation (1) (7). In order to simulate a large system on a computer we employ the usual periodic boundary conditions (2) i.e. a small central cube of length L, containing N particles is surrounded by periodic images. L is chosen to give the required volume fraction ϕ

$$L^3 = \frac{4\pi a^3 N}{3\phi} \times 100 \tag{3}$$

where a = particle radius.

PCS and Statistical Mechanics

In the absence of multiple scattering the electric field autocorrelation function $g(\underline{K},\tau)$ is related to the coherent intermediate scattering function $F(\underline{K},\tau)$ by[4]

$$g(\underline{K},\tau) = \frac{F(\underline{K},\tau)}{S(\underline{K})} , \tag{4}$$

where $S(\underline{K})$ is directly proportional to the angular distribution of the time averaged intensity of the scattered radiation and

$$K = \frac{4\pi n}{\lambda_o} \sin\frac{\theta}{2}$$

is the magnitude of the scattering vector, θ being the scattering angle. The intermediate scattering function can be expressed in terms of the positions of the scattering centers as[4]

$$F(\underline{K},\tau) = \frac{1}{N} \sum_{i=1}^{N} \sum_{j=1}^{N} \langle e^{i\underline{K}\cdot(\underline{r}_i/0)-\underline{r}_j(\tau))} \rangle , \tag{5}$$

where (for a stationary system) the angular bracket represents the ensemble average. Clearly $F(\underline{K},\tau)$ is simply the Fourier Transform of the van Hove space time correlation function

$$G(\underline{r},\tau) = \frac{1}{N} \langle \sum_{i,j}^{N} \delta(\underline{r} + \underline{r}_i(0) - \underline{r}_j(\tau)) \rangle \tag{6}$$

which can be split into the self $(i=j)$ and distinct $(i\neq j)$ space time correlation functions $G_s(\underline{r},\tau)$ and $G_d(\underline{r},\tau)$ respectively. Thus

$$F_s(\underline{K},\tau) = \int G_s(\underline{r},\tau)e^{-i\underline{K}\cdot\underline{r}} d\underline{r} \tag{7}$$

$$F_d(\underline{K},\tau) = \int G_d(\underline{r},\tau)e^{-i\underline{K}\cdot\underline{r}} d\underline{r} \tag{8}$$

where F_s, F_d are the self and distinct intermediate scattering functions. One may also define a time dependent self diffusion coefficient $D(\tau)$ by[5]

$$D(\tau) = 6 \int_o^\infty G_s(\underline{r},\tau) r^2 d\underline{r} \tag{9}$$

$$= \frac{<(\underline{r}_i(0) - \underline{r}_i(\tau))^2>}{6\tau} \qquad (10)$$

and

$$\lim_{\tau \to \infty} D(\tau) = D_s \qquad (11)$$

where D_s is the self diffusion constant measured by tracer
experiments (5). Thus one may use PCS to obtain G_s, G_d, $D(\tau)$ and
D_s. It has, unfortunately not been possible to date to measure
the incoherent intermediate scattering function to separate
$F(\underline{K},\tau)$ into its self and distinct components. This is where
neutron scattering (6) has, in principle, a distinct advantage.
Obviously, in the limit of extremely dilute dispersions
(independent particle diffusion) all the cross terms in equation
(5) vanish and $F = F_s$. Also when $K >> K_{max}$, where K_{max} is the
position of the primary maximum in $S(K)$, $F = F_s$ (4). A determination
of F_s in this way is, however, only possible when the modal
interparticle spacing is comparable with (or smaller than) the
wavelength of the radiation.

To date only very simple models have been used in an attempt
to extract this basic information from PCS data (4).

Application to Electrostatically Stabilized Colloids

A system which conforms to the assumptions inherent to the
BD method as outlined in the previous section is a very low ionic
strength aqueous suspension of charged polystyrene spheres.
Such systems have been extensively studied by PCS (4,8-10) since,
despite their low particle concentration (validity assumption (i)),
the particles are highly interacting because the Coulomb repulsion
between the charged surface groups is almost unscreened owing
to the very low ionic strength of the system. Thus they form
excellent models to study strongly interacting electrostatically
stabilized dispersions in the absence of collective hydrodynamic
effects.

The interaction potential for the N particles, Φ_N, we use is
the screened Coulomb repulsion, i.e.

$$\Phi_N = \sum_{i<j} U(r_{ij})$$

$$U(r_{ij}) = U(r_{ji}) = \frac{\pi \epsilon_o \epsilon_r 2a}{r} \psi_o^2 \exp \{-\kappa 2a(r-1)\} \qquad (12)$$

where a = particle radius, ϵ_r is the relative permittivity of
the dispersion medium, ϵ_o is the permittivity of free space, ψ_o

is the surface potential, κ is the Debye screening parameter given by

$$\kappa = \frac{2N_A ne^2}{\epsilon_o \epsilon_r kT} \qquad (13)$$

N_A being Avogadros number, n the elctrolyte concentration, e the electronic charge and r is expressed in particle diameters. Despite its simplicity the potential has been shown to give a reasonable description of the static structure of electrostatically stabilized aqueous polystyrene suspensions (11). Computations using this interaction potential were carried out using the following parameters; a = 23 nm, n = 10^{-3} mole m^{-3}, volume fractions, ϕ, in the range 0.015 to 0.070% and ψ_o in the range 90 to 220 mV. The free particle diffusion constant, D_o in equation (1) was obtained from the Stokes-Einstein relation, i.e.

$$D_o = \frac{kT}{6\pi\eta_o a} \qquad (14)$$

where η_o is the viscosity of the dispersion medium. All results were generated using N = 32, a time step of $\Delta t = 10^{-5}$ sec and a face centred cubic array of particles as a starting point. The first 1000 time steps were discarded to allow the system to equilibrate and a further 4500 time steps generated to simulate the dynamical behaviour of the system.

RESULTS AND DISCUSSION

From equation (1) we obtain for the N particles, a sequence in time of the positions of each particle at each of the M time steps

$$\{r_i(t_j) : i = 1, \ldots\ldots N; \; j = 1, \ldots\ldots M\}$$

From this sequence of positions we may calculate D(t) using (10), and the van Hove correlation functions using (12)

$$G_s(\underline{r},t) = \frac{1}{(4\pi r^2 \Delta r)} < N(\underline{r}_i(t) - \underline{r}_i(o) - \underline{r})>$$

$$G_d(\underline{r},t) = (\frac{1}{4\pi r^2 \Delta r}) < N(\underline{r}_j(t) - \underline{r}_i(o) - \underline{r})> \; i \neq j \qquad (16)$$

Figure 1 shows the mean square displacement of a particle (equation 10), for a range of volume fractions. These results are qualitatively very similar to that obtained experimentally by Pusey (4). It may be noticed that at short times, the particles

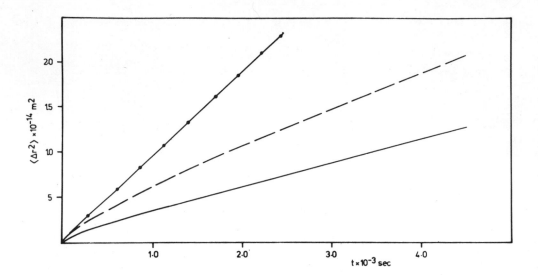

Fig. 1.   Mean square displacement versus time at ψ_o= 150 mV
          for; _____ · _____ · _____ free diffusion,

          ; ------------------- $\phi = 1.5 \times 10^{-4}$

          ; _____ $\phi = 7 \times 10^{-4}$

are undergoing essentially free diffusion, but as time increases
they start to feel the effects of their neighbours and their
motion is inhibited, and, in the long time limit, the self
diffusion coefficient has significantly decreased from its free
particle value ($D_0 = 9.55 \times 10^{-12}$ m^2 sec^{-1}). These effects
increase with an increase in volume fraction. Figure 2 shows
the variation of D(t) with surface potential. We see here the
large deviation from D_0 due to particle interactions. This
figure also shows an abrupt change in the diffusion coefficient
corresponding to the order-disorder transition, as predicted
by van Megen and Snook using the Monte Carlo method (11,13).
The behaviour of the van Hove self correlation function as a
function of ϕ may be seen in figure 3 for ψ_0 = 220 mV and t = 10^{-3}
sec. This is compared with the result of Brownian motion theory
for free particles (5) (solid line) i.e.

$$G_s(\underline{r},t) = \frac{1}{(4\pi D_o t)}3/2 \ \exp \ (-r^2/4D_o t) \tag{17}$$

and, as can be seen, as the volume fraction increases the effect
of the interactions tends to spread $G_s(\underline{r},t)$ less rapidly. The
van Hove distinct correlation function, which measures the
relaxation of the structure of the system with time is displayed

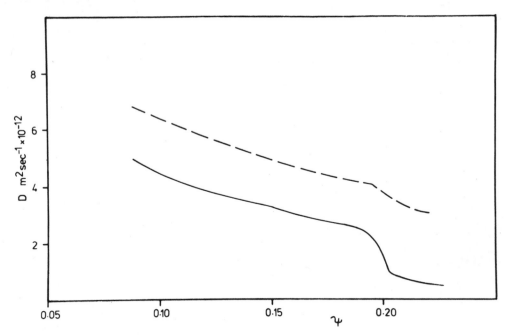

Fig. 2.   Diffusion coefficient versus surface potential (volts)
          at $t = 5 \times 10^{-4}$s _____; and

          $t = 5 \times 10^{-3}$s _ _ _ _ _ _ _ _ _ _ _ _ _ _ _ _ _ .

in figure 4 for $\phi = 3 \times 10^{-4}$% and $\psi_o = 220$ mV. As can be seen
$G_d(\underline{r},t) \rightarrow g(\underline{r})$ at very short times and $G_d(\underline{r},t) \rightarrow 1$ at very long
times.

CONCLUSION

        In view of the small number of particles and thus the
dimensions of the small periodic system in comparison with the
magnitude of the range of the interactions, the preceding
calculations are clearly preliminary. The results do however,
reproduce the features of experiments, suggesting that the
algorithm of Ermak is applicable to systems of strongly interacting spherical particles. Currently, these calculations are
being extended to larger numbers of particles and tested for
the long range part of the pair potentials. This will allow
a more realistic Fourier transformation of the van Hove
correlation functions so that the calculated results can be
compared directly with measured scattering functions rather
than extracted via models.

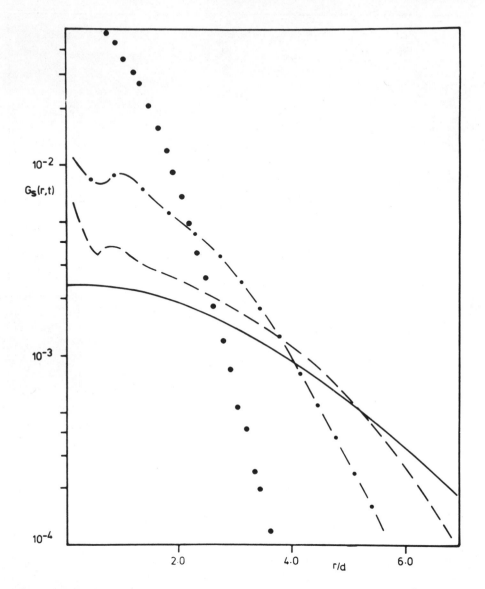

Fig. 3. van Hove self correlation function with $t = 10^{-3}$s and $\psi_o = 220$ mV.

_____ free diffusion;

- - - - - - - - - - - $\phi = 5 \times 10^{-5}$

— · — · — · — $\phi = 3 \times 10^{-4}$

· · · · · · · · · · · · · · · · $\phi = 7 \times 10^{-4}$

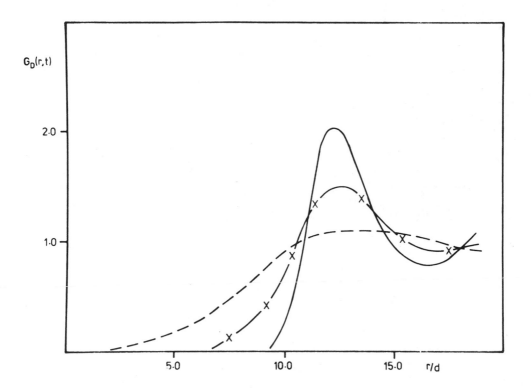

Fig. 4.   van Hove distinct correlation functions at
ψ_o = 220 mV and ϕ = 3 x 10^{-4}

————————————— t = 0

— + —— + —— t = 1 x 10^{-3}s

——————————— t = 4 x 10^{-3}s

REFERENCES

1. See for example, "Photon Correlation and Light Beating
    Spectroscopy", H. Z. Cummins and E. R. Pike, ed., NATO
    Advanced Study Institute Series Plenum, New York, (1974).
2. J. A. Barker and D. Henderson, Rev. Mod. Phys., 48:587 (1976).
3. D. L. Ermak and Y. Yeh, Chem. Phys. Letters, 24:243 (1974),
    D. L. Ermak, J. Chem. Phys., 62:4189, 4197 (1975).
4. P. N. Pusey, J. Phys. A., Math. Gen., 11:119 (1978).
5. R. O. Watts and I. J. McGee, "Liquid State Chemical Physics",
    Wiley-Interscience, New York, (1976).
6. S. H. Chen, in Physical Chemistry, An Advanced Treatise,
    H. Eyring, D. Henderson and W. Jost eds., Academic Press,
    New York, (1971), Vol. VIIIA.
7. D. L. Ermak and J. A. McCammon, J. Chem. Phys., 69:1352
    (1978).
8. J. C. Brown, P. N. Pusey, J. W. Goodwin and R. H. Ottewill,
    J. Phys. A., Math. Gen., 8:664 (1975).
9. D. W. Schaefer, J. Chem. Phys., 66:3980 (1977).
10. P. S. Dalberg, A. Bøe, K. A. Strand and T. Sikkeland,
    J. Chem. Phys., 69:5473 (1978).
11. W. van Megen and I. K. Snook, J. Chem. Phys., 68:813 (1977),
    Disc. Faraday Soc., 65:92 (1978).
    K. Gaylor, W. van Megen and I. K. Snook, JCS, Faraday II,
    75:451 (1979).
12. A. Rahman, Phys. Rev. A., 136:405 (1964).
13. I. K. Snook and W. van Megen, JCS Faraday II, 72:216 (1976),
    J. Colloid and Interface Sci., 57:40, 47 (1976).

LIGHT SCATTERING FROM STERICALLY STABILIZED SILICA PARTICLES

M.M. Kops-Werkhoven and H.M. Fijnaut

Rijksuniversiteit Utrecht
Van 't Hoff laboratorium voor Fysische en Colloïdchemie
Utrecht, The Netherlands

I. INTRODUCTION

During the last few years there has been considerable interest in the study of the dynamic properties of colloidal systems. This is mainly due to the development of the photon correlation spectroscopy, which makes it possible to study the Brownian motion of colloidal dispersions by measuring the particle diffusion coefficient. This experimental technique also stimulated much theoretical work on the influence of the interparticle interactions on the Brownian motion (1-14). The existing theories contain explicit results for the concentration dependence of the diffusion coefficient for interacting hard spheres in diluted systems, in the long wavelength limit.

Until now photon correlation experiments on interacting colloidal systems have been done mainly on charged spheres dispersed in a polar solvent (water) (15-24). Most of the experiments on these charges spheres have been done at low particle concentration under the condition of very low ionic strength, giving rise to electrostatic interactions between the particles of a range long compared with the particle dimensions. In this case the shorter ranged hydrodynamic interactions can be neglected. The particle concentration is kept low, in order to avoid multiple scattering. Because of the large interaction radii, compared to the wavelength of light, extrapolations to infinite wavelength are difficult for these measurements. Furthermore theories about the dynamics for these charged spheres are not sufficiently developed to compare theory and experiment in a quantitative way. Therefore in order to confirm the existing theories by experiments, measurements should be done on systems of small hard spheres, of which the range

of interaction between the particles is comparable with the
particle dimensions.

Recently some papers have been published, describing the
dynamics of systems of particles, of which the electrostatic interactions can be neglected (25, 26).

In our laboratory a system of particles has been developed
which appeared to behave as hard spheres as can be concluded from
static light scattering experiments (27). The particles consist of
a nucleus of silicon dioxide (SiO_2) surrounded by a spherical shell
of hydrocarbon chains. Because of its small dimensions compared
with the wavelength of light and its hard sphere character, the
particle seems to be very well suited to test the existing hydrodynamic theories. The measurements should be done at relatively
high concentrations. In order to minimize the effect of multiple
scattering the particles are dispersed in a solvent of which the
index of refraction is close to that of the silica particles.

In sections II and III a description of the preparation and
properties of the silica particles will be given. In the two
following sections IV and V, the basic light scattering equations,
the experimental equipment and the results of the theories
developed for the concentration dependence of the diffusion
coefficient for hard spheres will be mentioned. The results of
the light scattering experiments carried out on the system of
silica particles dispersed in cyclohexane will be described in
section VI. In the last section VII the conclusions which can be
drawn from the measurements are discussed.

II.   PREPARATION OF THE SILICA PARTICLES

The preparation of the silica particles was first mentioned in
the literature by Stöber (28). A further development and
characterization of these particles was done in our laboratory by
van Helden (29). Briefly, the procedure of preparation of the
silica particles is as follows.

Silica particles were made by a condensation reaction of
silicic acid in ethanol, which gives under certain specific
conditions spherically symmetric particles. In order to stabilize
the particles sterically, a reaction of the hydrozyl groups on the
surface of the particle is performed with an alcohol, in this case
stearylalcohol, by means of an etherification (3). After purification and drying, the silica particles can be dispersed in apolar
solvents. By changing the initial conditions during the condensation
reaction, the radius of the particles can be varied between 20 and
100 nm.   The particles used in the following experiments had a
radius of 22-23 nm.

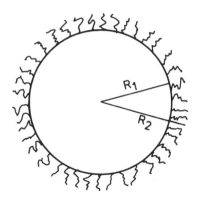

Fig. 1. A sterically stabilized particle. The nucleus consists
        of silicondioxide (SiO$_2$) and has a radius of 20 nm (R$_1$).
        The surrounding layer consists of hydrocarbon chains (c-18).
        The particle radius (R$_2$) is 22-23 nm.

III. CHARACTERIZATION OF THE SILICA PARTICLES

     The silica particles used in the experiments were characterized
by static and dynamic light scattering and electron microscopy.
The static light scattering experiments showed that the radius of
the particles after the condensation reaction in ethanol was 20 nm.
After etherification the radius of the particles was expected to
be 22-23 nm, because of the length of the hydrocarbon chains on the
surface of the particle. Static light scattering experiments on
these sterically stabilized silica particles will not directly
give an exact particle radius, because of the difference in index of
refraction of nucleus and surface layer (32). The radius of
gyration measured with static light scattering is an optical radius.
When the difference in the index of refraction of the solvent and
the particle is small,optical fluctuations in the particle itself
will give a change in the radius of gyration. With dynamic light
scattering however, the hydrodynamic radius of the particles in a
very diluted solution appeared to be 23 nm, which was in agreement
with the expected particle dimension. Changing the solvent had no
measurable effect on the hydrodynamic radius of the particles.

     Electron microscopy pictures showed that the particles were
spherical. The standard deviation of the particle radius found
from these pictures was about 10%.

     In order to minimize the effect of multiple scattering, the
particles were dispersed in a solvent, cyclohexane, of which the
index of refraction (n$_s$ = 1.425) was about the same as the mean
index of refraction of the silica particles ($\bar{n}_p$ = 1.44).

An important step during the sample preparation was the sonication of the particles in order to disperse them in apolar solvents. The solutions were sonicated in a water bath during one hour, after which the solutions remained stable and no further changes in the particle radius was measured.

The solvent cyclohexane was made dust free by filtration through a millipore filter (pore diameter 25 nm). In order to remove the remaining dust from the silica dispersions, the samples were centrifuged during 5 minutes at 5000 r.p.m. in an ultracentrifuge.

IV. LIGHT SCATTERING EQUATIONS AND EXPERIMENTAL EQUIPMENT

The static light scattering experiments were done with a FICA 50 apparatus which was used at two different wavelengths: 546 nm and 436 nm. The average of the scattered intensity $I_s(K)$ as a function of wave vector K is given by

$$<I_s(K)> = A \, I_o \, c \, P(K) \, S(K) \tag{1}$$

where the brackets denote the ensemble or time average for the intensity that is assumed to be stationary. The relation between K and the angle of observation θ is

$$K = \frac{4\pi}{\lambda} \sin \frac{\theta}{2} \tag{2}$$

where λ is the wavelength of the incident light in the scattering system. A is a constant, I_o the intensity of the incident beam, c the particle concentration (gram/cm^3) and P(K) is the interparticle interference factor. The time average structure factor (S(K), representing the interparticle interference is defined by:

$$S(K) = 1 + 4\pi\rho \int_o^\infty r^2\{g(r)-1\} \frac{\sin Kr}{Kr} \, dr \tag{3}$$

Here ρ is the mean number density and g(r) the radial distribution function, which gives the probability density of finding a particle at a distance r from another particle at r = 0.

In the limit of low density S(K) = 1

The dynamic light scattering experiments were done with a Honeywell Saicor 42 A, 100 points correlator, used in the photon counting mode and a Brül and Kjaer type 2114 spectrum analyser. This spectrum analyser was modified for a parallel output of 50 channels.

By measuring the time autocorrelation function of the scattered intensity or by measuring the spectral density the dynamics of the particles can be examined. The homodyne autocorrelation function at low particle density can be written as:

$$g(K,\tau) = \frac{<I_s(K,t)I_s(K,t+\tau) - <I_s(K,t)I_s(K,t)>}{<I_s(K,t)I_s(K,t)>} = \exp(2D_{eff}K^2\tau)$$

$$(4)$$

where D_{eff} is the effective diffusion coefficient. At infinite dilution D is the Stokes-Einstein diffusion coefficient D_o, which is for spherically symmetric particles:

$$D_o = \frac{k_B T}{6\pi\eta a_h}$$

$$(5)$$

where k_B is the Boltzman constant, T is the temperature, η the solvent viscosity and a_h the hydrodynamic particle radius.

The normalized homodyne power spectrum measured with the spectrum analyser is given by

$$I(K,\omega) = \frac{2 K^2 D}{\omega^2 + (2K^2D)^2}$$

$$(6)$$

where ω is the angular frequency.

V.   PREDICTED THEORETICAL BEHAVIOUR OF THE EFFECTIVE DIFFUSION COEFFICIENT FOR HARD SPHERES

In the last few years several theories were developed (1.14) describing the concentration dependence of the effective diffusion coefficient. Explicit results are given for hard spheres in diluted systems mainly in the long wavelength limit (K→0).

The relation between D_{eff} and the volume fraction ϕ of the particles in solution can be given by the following equation:

$$D_{eff}(K \to 0) = D_o(1+\alpha\phi)$$

$$(7)$$

where α is a constant.for hard spheres.

From the developed theories, different values of α were obtained (table I).

VI.   RESULTS

The static light scattering results from samples of silica particles in cyclohexane with a concentration (c) up to 0.17 g cm^{-3} are shown in figure 2.

In this concentration range the dependence of the inverse of the scattered intensity $I_s^{-1}(K)$ on K^2 is still linear, as could

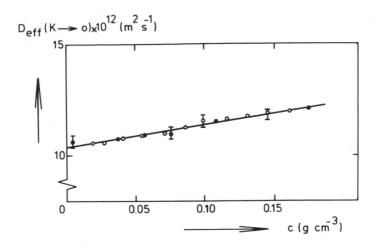

Fig. 3. Dependence of the effective diffusion coefficient
extrapolated to K = 0 on the concentration of silica
particles c.

$$D_o = (10.4 \pm 0.1) \times 10^{-12} \ m^2 s^{-1}$$

o and ● correspond to a dilution series of the samples
and to some directly prepared samples respectively.

VII. CONCLUDING REMARKS

The following conclusions can be drawn from the measurements.
First we can see that the concentration dependence of the effective
diffusion coefficient extrapolated to K = 0 can be described as a
linear function in the observed concentration range and within the
experimental error.

The value of α found from the experiments is between 12 and
16. This range includes the theoretical values of Batchelor (5)
(α = 1.45) and Felderhof (14) (α = 1.56).

The static light scattering experiments carried out on the
same system already showed that the behaviour of the silica
particles can be described by a system of hard spheres (27).

The dynamic light scattering experiments on this system
also give a good agreement with the theories of hard spheres in
solution. Further measurements on this system will be done,
especially measurements of the sedimentation coefficient as a
function of the particle concentration, measurements in other
solvents and at higher concentration range.

ACKNOWLEDGEMENT

This work was part of the research programme of the "Strickting voor Fundamenteel Onderzoek der Materie" (F.O.M.) with financial support from the "Nederlandse Organisatie voor Zuiver-Wetenschappelijk Onderzoek" (Z.W.O.).

References

1. J. M. Burgers, Proc. K. Ned. Akad. Wet. Amsterdam, 44,1047, 1177 (1941). Proc. K. Ned. Akad. Wet. Amsterdam, 45, 9, 126 (1952).
2. C. W. Pyun and M. Fixman, J. Chem. Phys. 41:937 (1964).
3. R. Zwanzig, Adv. Chem. Phys., 15:325 (1969).
4. A. M. Kapustin, T. N. Khazanovich, Coll. J. USSR, 31:675 (1969). Kolloidn. Zn., 31:842 (1969).
5. G. K. Batchelor, J. Fluid. Mech., 52:245 (1972). J. Fluid. Mech.
6. A. R. Altenberger and J. M. Deutch, J. Chem. Phys., 59:894 (1973).
7. J. L. Aguirre and T. J. Murphy, J. Chem. Phys., 59:1833 (1973).
8. G. D. Phillies, J. Chem. Phys., 50:976, 978 (1974). J. Chem. Phys., 62:3925 (1975).
9. S. Harris, J. Phys. A: Math. Gen., 9:1895 (1975).
10. J. L. Anderson and C. C. Reed, J. Chem. Phys., 64:3240 (1976).
11. A. R. Altenberger, Chem. Phys., 15:242 (1976). Physica A92:391 (1978). J. Chem. Phys., 70:1994 (1979).
12. B. J. Ackerson, J. Chem. Phys., 64:242 (1976).
13. W. Hess and R. Klein, Physica A85:509 (1976).
14. B. U. Felderhof, J. Phys. A: Math. Gen., 11:929 (1978).
15. P. N. Pusey, J. M. Vaughan, G. Williams, J. Chem. Soc. Far. Trans. II 70:1696 (1974).
16. J. C. Brown, P. N. Pusey, J. W. Goodwin, R. H. Ottewill, J. Phys. A: Math. Gen., 8:664 (1975).
17. J. L. Anderson, F. Rauh, A. Morales, J. Phys. Chem., 82:608 (1978).
18. P. N. Pusey, J. Phys. A: Math. Gen., 11:119 (1978).
19. A. P. Minton and P. D. Ross, J. Phys. Chem., 82:1934 (1978).
20. B. D. Fair, D. Y. Chao, A. M. Jamieson, J. Colloid Interface Sci., 66:323 (1978).
21. D. R. Bauer "Hydrodynamic properties of dilute and concentrated polymer latexes: A study by quasi-elastic light scattering", Engineering and Research Staff, Ford Motor Company Dearborn, Michigan 481211 (1978).
22. G. Jones and D. Caroline, Chem. Phys., 37:187 (1979).
23. W. Burchard and M. Schmidt, Ber. Bunsenges. Phys. Chem., 83:833 (1979).
24. L. O. Sundelöf, Ber Bunsenges. Phys. Chem., 83:329 (1979).

25. H. M. Fijnaut, C. Pathmamanoharan, E. A. Nieuwenhuis, A. Vrij,
     Chem. Phys. Letters, 59:351 (1978).
26. A. M. Cazabat, D. Langevin, A. Pouchelon "Light scattering
     study of W/O emulsions" Laboratoire de Spectroscopie
     Hertzienne de l'E.N.S. 24, rue Lhomond, 75231 Paris Cedex
     05. Workshop of quasi-elastic light scattering studies,
     Milan (1979).
27. A. K. van Helden and A. Vrij, to be published in J. Colloid
     Interface Sci.
28. W. Stöber, A. Fink, E. Bohn, J. Colloid Interface Sci.,
     26:62 (1968).
29. A. K. van Helden, J. W. Jansen and A. Vrij, to be published in
     J. Colloid Interface Sci.
30. R. K. Iler, U.S. Patent 2, 801, 185.
31. P. N. Pusey, D. E. Koppel, D. W. Schaefer, R. D. Camerini-
     Otero, S. H. Koenig, Biochemistry 13:952 (1974).
32. A. K. van Helden and A. Vrij, to be published in J. Colloid
     Interface Sci.

DEPOLARIZED LIGHT SCATTERING FROM CONCENTRATED PARTICLE

SUSPENSIONS*

Brigitte Herpigny and Jean Pierre Boon

Faculté des Sciences CP 231
Université Libre de Bruxelles
B-1050- Bruxelles, Belgium

ABSTRACT

We present preliminary measurements of the static and dynamic structure factors of concentrated suspensions of spherical polystyrene particles by intensity correlation spectroscopy. Polarized and depolarized light scattering experiments were performed as a function of concentration and of scattering angle for particles with diameter a $\simeq$ 1μm and concentrations n* = Na^3/V between 4×10^{-4} and 2×10^{-2}. The correlation time is found to be independent of scattering angle and decreases with increasing concentration. The shpae of the total scattering intensity is quite reminiscent of the static structure factor obtained by neutron scattering for simple dense fluids.

MOTIVATION

Light scattering studies of suspensions of interacting particles with diameter of the order of the wavelength of the incident light have shown that such systems exhibit a structure similar to that of a dense fluid as observed by neutron scattering (1,2). On the other hand, depolarized light scattering spectra have been measured for Brownian systems and the results have been interpreted in terms of multiple scattering (3,4). Depolarized light scattering has also been studied in simple monatomic fluids and the spectra were interpreted as resulting from the effects

*Work supported in part by the Fonds National de la Recherche Scientifique (FNRS, Belgium).

of pair interactions (5). Pursuing the line of reasoning that
emerges from the combination of the studies cited above, one is
led to consider whether collision-induced scattering could be
observed in concentrated suspensions of large particles by
depolarized light scattering.

EXPERIMENTAL ASPECTS

We have performed light scattering experiments on polystyrene
spheres with diameter a = 1.0910 ± 0.0082 μm (Dow-Latex 41951)
suspended in ethanol at various concentrations ranging from
4×10^{-4} to 2×10^{-2} when expressed in reduced units: $n^* = na^3 = V_o/V$,
where V_o is the close-packed volume. The measurements were made
by intensity correlation spectroscopy with standard set-up in the
homodyne detection mode at scattering angles of 20°, 30°, 40°, 50°.
A typical spectrum obtained in the VH geometry is shown in Fig. 1.

Total intensity measurements were also performed for each
concentration over the whole angular range (0° - 180°) by means of
a photodiode cell.

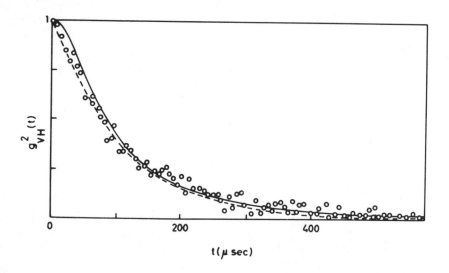

Fig. 1.   Normalized depolarized light scattering spectrum of
          suspended particles in ethanol; diameter: a = 1.091 μm;
          concentration $n^* = Na^3/V = 10^{-2}$; scattering angle:
          θ = 20°; scattering geometry: VH. The solid curve and
          dashed line are obtained by fitting the experimental data
          (open circles) to a Lorentzian and an exponential
          respectively.

RESULTS AND DISCUSSION

Two causes of depolarization of the light scattered by dense
systems of spherical particles are (i) multiple (mainly double)
scattering, and (ii) colision-induced (or interparticle) scattering.
It is inferred from earlier studies (3,4,6) that double scattering
spectra are well represented by a function decaying exponentially
in time, whereas intermolecular scattering spectra exhibit an
exponential decay in frequency (5). On the basis of the assumption
that in concentrated suspensions of spherical particles double
scattering and interparticle scattering can cause depolarization
of the scattered light, one is thus led to attempt one parameter
fits of the spectra with the functions $f_e(t) = \exp(-t/\tau)$ and
$f_L(t) = \{1 + (t/\tau)^2\}^{-1}$. As seen from fig. 1, it appears that the
initial decay is better described by an exponential function
whereas at longer times a Lorentzian is more appropriate.
Apart from a constant factor, both functions yield the same
correlation time $\tau_c = \int_0^\infty dt\, f(t)$. So, it is of no importance which
function is used to derive τ_c from the experimental spectra to
study the correlation time as a function of scattering angle and
concentration.

Figure 2 shows that τ_c is essentially k-independent (k =
$4\pi\, n_0 \lambda_0^{-1} \sin \theta/2$) for a concentrated suspension. Also shown is
the slope $2D_0$, where D_0 is the Brownian diffusion coefficient, as
obtained from unpolarized scattering experiments performed in
dilute suspensions of the same polystyrene spheres. For the
concentrated suspension (here $n^* \simeq 10^{-2}$) the free diffusion regime
would be reached for $ka \geq 100$, a value not accessible in this
particular experiment.

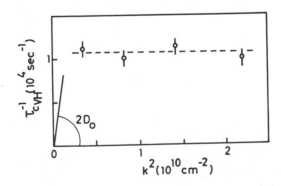

Fig. 2.   The reciprocal correlation time as a function of the
          scattering wavenumber for a suspension at $n^* = 10^{-2}$. D_0
          is the diffusion coefficient of the corresponding
          Brownian system..

The concentration dependence of the correlation time shows a rapid decay of τ_c in the low concentration range followed by an asymptotic approach towards a constant value at high concentrations. A logarithmic plot of the variation of τ_c is presented in Figure 3 indicating a power law $\tau_o/\tau_c \propto \sqrt{n^*}$ with $\tau_o^{-1} = 2D_ok^2$; thus $\tau_c^{-1} \propto n^{*1/2}\, T$. This cannot be explained on the basis of an analogy with simple dense fluids for which the second moment of the velocity autocorrelation function yields $\Omega_o \propto n^*\, T^{1/4}$.

Assuming that the characteristic time measured by depolarized scattering corresponds to the duration of the correlation of two particles diffusing towards each other, one can write $<x^2> = 2D_o\tau_c$ where x is the projected distance over which the particles are in interaction. For particles with diameter $a \simeq 1\mu m$ and $n^* \simeq 10^{-2}$, $\tau_c \sim 10^2 \mu s$. Then, the estimate yields $\sqrt{<x^2>} \sim 50\text{Å} << a$ which would indicate that the suspension can be interpreted as a "hard-sphere fluid."

One would expect that at concentrations such that $10^{-4} \lesssim n^* \lesssim n^{-2}$, the suspension exhibits a hard-sphere liquid like structure. The static structure can be obtained by total intensity measurements as a function of the scattering angle (1,8). The experimental results are presented in Figure 4. The static structure measured by unpolarized light is quite similar to S(k) in a dense monatomic fluid but the position of the first diffraction peak corresponds to a value of ka considerably lower than in a hard-sphere liquid.

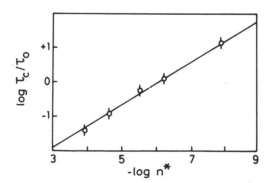

Fig. 3.   The correlation time (obtained from spectra measured at $\theta = 20^o$) as a function of the concentration. $\tau_o = (2D_ok^2)^{-1}$. The slope of the solid line corresponds to the power law $\tau_c^{-1} \propto n^{*1/2}$.

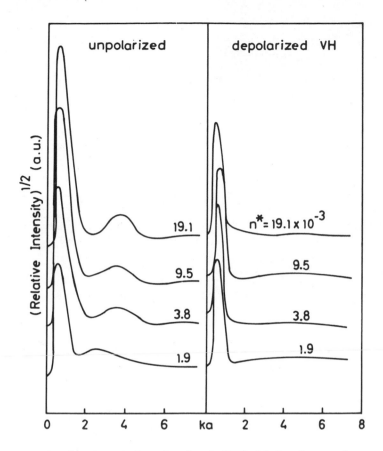

Fig. 4. Unpolarized and depolarized (VH) light intensity as a
         function of wavenumber for various concentrations.

     The second diffraction peak becomes less and less pronounced
when lowering the concentration as is the case in a hard-sphere
fluid when the density decreases (9). On the other hand, we
observe no second diffraction peak in the depolarized light
intensity measurements. It is quite well known that despite the
one-to-one correspondence between S(k) and the radial distribution
function g(r), the interpretation of the structure of S(k) in
terms of first neighbour distances is often ambiguous. In the
present case it remains to be determined whether the lack of
second diffraction peak in the depolarized light intensity
measurements is, or is not, an indication that first neighbour
interactions are strongly dominant and consequently that depolarized
light scattering would probe mainly the dynamics of pair collisions.

Acknowledgements

One of us (B.H.) acknowledges the benefit of a grant from the Institut pour l'Encouragement à la Recherche Scientifique dans l'Industrie et l'Angriculture (IRSIA, Belgium).

References

1. J. C. Brown, P. N. Pusey J. W. Goodwin, and R. M. Ottewill, J.Phys.A., 8:664 (1975).
2. P. N. Pusey, J.Phys.A., 11:119 (1978).
3. C. M. Sorensen, R. C. Mockler, and O'Sullivan, Phys. Rev. A., 14:1520 (1976).
4. J. Boheim, W. Hess, R. Klein, Z. Physik B., 32:237 (1979).
5. P. A. Fleury, W. B. Daniels, and J. M. Worlock, Phys. Rev. Lett. 27:1493 (1971).
6. P. C. Colby, L. M. Narducci, and J. Baer, Phys. Rev. A., 12:1530 (1975).
7. J. W. E. Lewis and S. W. Lovesey, J. Phys. C., 10:3221 (1977).
8. D. W. Schaefer, J. Chem. Phys., 66:3980 (1977).
9. N. W. Ashcroft and J. Lekner, Phys. Rev., 145:83 (1966).

INTRAMOLECULAR MOTION OF POLYSTYRENE

David Caroline and Gwynne Jones

School of Physical and Molecular Sciences
University College of North Wales
Bangor, Gwynedd, U.K.

INTRODUCTION

Intramolecular motion in macromolecules of high molecular weight($\bar{M}_w > 10^6$) in dilute solutions has been investigated using quasi-elastic light-scattering techniques by a number of experimenters, and most (1-4) have interpreted their results in terms of Pecora's analysis (5-7) of the normal modes of motion of the Rouse-Zimm bead-and-spring model.

We have used photon-correlation spectroscopy to carry out a systematic investigation of intramolecular motion in polystyrene in two theta solvents (8,9), firstly cyclohexane at $34.5^\circ C$ and secondly a mixed solvent (80.25 vol % carbon tetrachloride/methanol at $25^\circ C$) which was established as a theta solvent in a previous investigation (10). Five narrow fractions were studied with $\bar{M}_w/\bar{M}_n$ between 1.02 and 1.14. For each polymer sample, dilute solutions were prepared at up to six different concentrations, well below overlap, in both solvent systems. For each solution, correlation functions of the scattered light were obtained over a range of angles from 13° to 120°, and these were analysed to obtain τ_1, the relaxation time of the first normal mode of intramolecular motion. The values obtained are up to twice those reported elsewhere, but agree well with the predictions of Zimm's (11) non-draining bead-and-spring model, modified for the light-scattering case.

At sufficiently low forward scattering angles, the macro-molecules behave as point scatterers, and measurements simply yield a value for the translational diffusion coefficient D. This will be the case when the parameter x is much less than unity; $x = K^2 \langle S^2 \rangle$, where K is the scattering vector and $\langle S^2 \rangle$ the mean-

square radius of gyration. The value of D obtained is used in
the analysis of data at higher angles, where x>1, in order to
extract a value for τ_1, the first of a series of relaxation times
τ_j each associated with the corresponding normal mode. For
example, in photon-correlation spectroscopy, the time-dependent
part of the intensity correlation function $G(\tau)$ is a single-
exponential function at low angles (x<<1) with decay rate $2DK^2$,
but turns into a multi-exponential function at higher angles with
the τ_j's appearing in the decay times. Up to the present, it has
proved possible only to obtain values for τ_1.

Zimm (11) has derived an expression for relaxation times
that are appropriate to dynamic viscosity measurements; these
refer to the relaxation of the ensemble average of the square of
the normal coordinate. However, Schurr (12) has pointed out that
polarized dynamic light-scattering experiments measure the
Langevin relaxation times associated with the relaxation of the
normal coordinate rather than its square. Fortunately, the
light-scattering times are simply twice the corresponding visco-
elastic times, so Zimm's expression can be adapted for the light-
scattering case to give

$$\tau_j = \frac{M \eta_o \eta}{A_j RT} \tag{1}$$

where M is the polymer molecular weight, η_o the solvent viscosity,
$[\eta]$ the intrinsic viscosity of the solution, R the gas constant
and T the absolute temperature. A_j equals $\pi^2 j^2/12$ in the Rouse (13)
free-draining case, but in the non-draining case associated with
strong hydrodynamic interaction it equals $0.293\lambda'_j$, where λ'_j
is the jth reduced eigenvalue of the matrix HA; H is the hydrodynamic interaction matrix and A the matrix associated with the
spring force of the polymer chain. Using A in its pre-averaged
form, Zimm (11) obtained a value of 4.04 for λ'_1, whereas in a
recent calculation which avoids pre-averaging (14) Bixon and
Zwanzig have obtained a value of 1.96. Thus, although Bixon
and Zwanzig's estimate of τ_1 is twice Zimm's, in both cases
τ_1 should vary as $M^{3/2}$ since $[\eta]$ varies as $M^{1/2}$ under theta
conditions; this contrasts with the free-draining case of Rouse
where M^2 dependence is predicted for τ_1.

The normal-mode analysis scheme of Pecora is reliable
experimentally only over a small range of x above unity. As x
increases, the higher relaxation times play a greater role
making it virtually impossible to analyse the data in terms of the
individual τ_j's. To overcome this problem, Saleh and Hendrix (15)
have calculated the form of the correlation function $G(\tau)$ in
the Rouse free-draining limit for values of x up to 50, but
their experimental study (16) did not support the results for this

model. This approach has been extended by Freire (17) to the
more realistic non-draining model of Zimm. Since the principal
effect of hydrodynamic interaction is to change the values of D
and τ_j, he was able to use the simpler free-draining for the
elements of the matrix Q (associated with the transformation to
normal corordinates) and for $\langle\mu_j^2\rangle e$(the mean-squared equilibrium
length of the jth normal mode coordinate). The results of
Freire's study are in closer agreement with the experimental
results of Hendrix et al (16) than the free-draining calculation.

    Büldt (18) has proposed a simpler scheme of data analysis
in which the normalized initial slope of the time-dependent
part of the experimentally determined $G(\tau)$ is plotted as a
function of x, and compared with theoretical predictions obtained
from calculations of the dynamic structure factor. His calculation
of the initial slope was for the free-draining case; the non-
draining case is covered in Freire's work.

    Schurr (12) has examined the form of the correlation function
for the Rouse-Zimm bead-and-spring model, and concluded that, at
sufficiently high values of x, its time dependence can be well
represented by a single exponential characterizing the diffusive
motion of individual beads. However, very large values of x are
required since the scattering wavelength $\Lambda(=2\pi/K)$ must be smaller
than the bead radius r; the reduced hydrodynamic interaction
$h^*(=r/l$ where l is the bead separation) is thus less than unity.
This condition requires $x > N/(h^*)^2$, where N is the number of
beads in the chain, and may not be reliable in light-scattering
experiments.

    Using a quite different approach that employed the
projection-operator technique, Akcasu and Gurol (19) had previously
reached the same conclusion as Schurr for high x values. In fact
they were able to distinguish three separate x regions, with an
intermediate region ($1 << x << N/6$) separating the low and high x
regions. An apparent diffusion coefficient D_{app} is obtained
from the initial slope of the correlation function. As expected
D_{app} is constant and equal to D in the low x region, where the
correlation function is completely dominated by the translational
diffusive motion of the coil. D_{app} is also constant in the high
x region where the diffusive motion of the single beads dominates.
However, in the intermediate region, D_{app} is expected to vary
linearly with K as the number of beads within Λ decreases.
Burchard (20) has recently extended this approach to deal with
polydisperse samples, and obtained a relationship between
$(D_{app}(x)/D_z)$ and x, where D_z is the value of D_{app} as $x \to 0$, the
z-average value of D.

EXPERIMENTAL

The narrow fractions of high molecular weight polystyrene were obtained from the Toyo Soda Manufacturing Co. of Japan, and the solutions were made using BDH Analar or spectroscopic grade reagents. The solutions were clarified by filtration through 0.5 or 1 μm Millipore filters directly into glass spectroscopic cells of square section and 10 mm path length. They were housed on the spectrometer table in an enclosure which maintained the temperature constant to within $0.05^{\circ}C$. The enclosure was mounted on an adjustable x-y stage to enable the cell to be moved to suit the angle of viewing.

No index-matching fluid was used, but the reflectivity of the exit window of the cell was reduced from 4% to 0.3% by fixing onto it, with microscope oil, a glass slide with a double-layer anti-reflection coating on its outside surface. At forward angles, observations of the scattered light were made through the exit window without the anti-reflection plate since the back-scattering from the reflected beam was very weak. Tests showed no difference with the plate in position, and the single-exponential nature of the correlation functions demonstrated the absence of any detectable contribution from the reflected beam. However, it was essential to keep the cell extremely clean and prevent the build-up of dust on the exit window, which could act as a local oscillator for heterodyning.

The beam from a Coherent Radiation Model CR2 argon-ion laser operating at 488 nm was focused into the solutions, and the scattered photons were detected by an ITT FW 130 photomultiplier tube. The resulting pulses were fed into a 48-channel digital autocorrelator (Malvern Instruments K7023) which computed the clipped autocorrelation function of the scattered intensity.

Correlation Functions

Pecora's theoretical study of the spectrum of the light scattered by a solution of large macromolecules in terms of the normal modes of motion gives a normalized first-order (optical) correlation function $g(1)(\tau)$ as a series of exponentials:

$$|g^{(1)}(\tau)| = e^{-DK^2\tau} (P_o + P_{21}e^{-2\tau/\tau_1} + P_{12}e^{-\tau/\tau_2}$$
$$+ P_{22}e^{-2\tau/\tau_2} + P_4 e^{-4\tau/\tau_1} + \dots)$$

The values of the coefficients P depend strongly on x, but for x less than 2.5, only the first two terms inside the brackets are significant, with P_o dominating. However, as x increases, the other coefficients assume an increasing importance and a complex expression results.

The normalized second-order (intensity) correlation function can be written down using the Siegert relation,

$$g^{(2)}(\tau) = 1 + |g^{(1)}(\tau)|^2. \quad \text{Thus}$$

$$g^{(2)}(\tau) - 1 = P_o^2 \, e^{-2DK^2\tau} \left[1 + \frac{2P_{21}}{P_o} e^{-2\tau/\tau_1} + \left(\frac{P^2_{21}}{P^2_o} + \frac{2P_4}{P_o} \right) e^{-4\tau/\tau_1} \right.$$

$$\left. + \frac{2P_{12}}{P_o} e^{-\tau/\tau_2} + \frac{2P_{22}}{P_o} e^{-2\tau/\tau_2} + \cdots \right] \qquad (2)$$

Only the leading terms have been included, and for small x this reduces to

$$g^{(2)}(\tau) - 1 = P_o^2 \, e^{-2DK^2\tau} \left(1 + \frac{P_{21}}{P_o} e^{-2\tau/\tau_1} \right)^2 ,$$

since the coefficient $2P_4/P_o$ in the previous equation is much smaller than $(P_{21}/P_o)^2$.

At forward angles, the time dependent part of the intensity correlation functions $G(\tau)$ was found to be a single exponential as expected theoretically for x << 1. The span of the trace was chosen to be approximately four decay times to optimise the efficiency of the curve-fitting procedure. The data from the correlator were first divided by the normalization constant supplied by the correlator, and this normalized data $g(\tau)$ was then computer-fitted to a function $1 + \alpha \exp(-\Gamma\tau) + \delta$, with α, Γ and δ as adjustable parameters. Although, ideally, δ should be zero, it has been found (10) necessary to include this misnormalization term to obtain consistently reproducible values of Γ; δ fluctuates from run to run and is typically less than one per cent of α.

The values obtained for D at low angles ($\Gamma = 2DK^2$) showed no systematic variation with angle and agreed within 1% over a wide range of angle. An example of the single exponential computer fit of $g(\tau)$ is shown in figure 1; more significantly, the difference between the experimental points and the fitted curve is displayed, and these residuals, magnified five times, are seen to be random. (The initial data point is ignored in the fitting procedure since it is usually affected by after-pulsing in the photomultiplier tube).

At higher angles, the multi-exponential nature of $g(\tau)$ becomes apparent if a single exponential is force-fitted to the data, as seen in figure 2; and the residuals now take on a characteristic pattern. The normalized data were fitted to the function

$$1 + e^{-2DK^2\tau} (a + be^{-\gamma\tau})^2 + \delta,$$

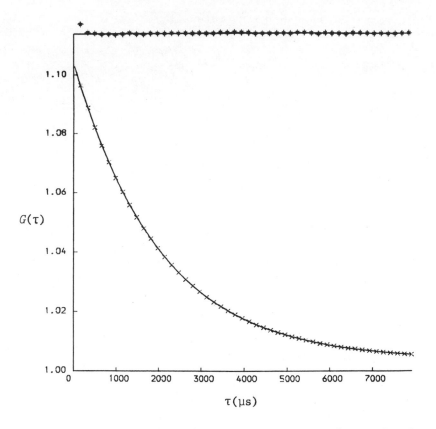

Fig. 1.   Single exponential fitted to the data points of a low-
          angle correlation function ($\bar{M}_W$ = 2.88 x 10^6 in cyclohexane
          at 34.5°C; 0.80 mg/ml, θ = 18.60°). The difference
          (x5) between the data points and the fitted curve is also
          shown.

with a, b, γ and δ as adjustable parameters; D was supplied from
the low-angle results. With this four-parameter fit, the
residuals are random.

     As θ increases, and values of.x increase beyond two, the
four parameter fit will not be such a good approximation to
the theoretical expression of equation 2 since increasingly
significant terms are being neglected (15). At these higher x
values, a force-fitted value for γ will yield a 'collective'
relaxation time τ_c (equal to 2/γ) which might be expected to
decrease with x since the extra exponentials have decay times
shorter than the first. This was found to be so experimentally,
but for values of x between 1 and 2,τ_c was found to be constant,
and this was taken to be the value of τ_1.

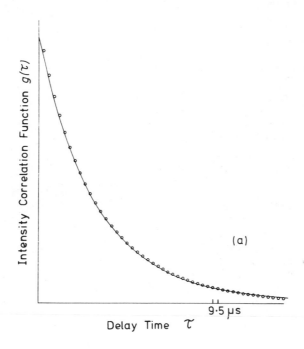

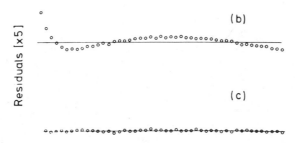

Fig. 2. (a) Single exponential force-fitted to the data points of a high-angle correlation function ($\bar{M}_w$ = 2.88 x 10^6 in cyclohexane at 34.5°_c, 1.15 mg/ml, θ = 83°).

(b) Difference (x5) between the data points and the fitted curve.

(c) Residuals (x5) for four-parameter fit of data in (a).

RESULTS AND DISCUSSION

Diffusion

In all cases, D was found to decrease linearly with polymer concentration, and in both solvent systems the value of D at infinite dilution, D_o, varied as $M_w^{\frac{1}{2}}$, as expected at the theta point. For the cyclohexane system the results are best represented by the expression,

$$D_o = (1.46 \pm 0.2) \times 10^{-4} \, \bar{M}_w^{-(0.508\pm0.007)} \, cm^2 s^{-1},$$

while in the mixed-solvent system

$$D_o = (1.10 \pm 0.04) \times 10^{-4} \bar{M}_w^{-(0.499\pm0.003)} \, cm^2 s^{-1}.$$

A mean hydrodynamic radius R_h can be defined from the Stokes-Einstein equation, $D = kT/6\pi\eta_o R_h$, where k is Boltzmann's constant . For each narrow fraction, the values obtained for R_h in the single and mixed solvent systems agreed within 1%, and are represented by the equation $R_h = 0.228 M_w^{\frac{1}{2}}$ Å.

In recent light-scattering measurements, Miyaki et al (21) found that $<S^2>^{\frac{1}{2}} = 0.297 \bar{M}_w^{\frac{1}{2}}$ Å for high $\bar{M}_w$ polystyrene in cyclohexane at the theta point. Thus the experimental value for the ratio $R_h/<S^2>^{\frac{1}{2}}$ is 0.77, which is 16% greater than the theoretical ratio of 0.66 for a non-draining random coil (22). This discrepancy in the friction coefficient contrasts with the very good agreement between experiment and theory for the intrinsic viscosity of theta systems (23).

Relaxation Times

For each solution, about ten sets of data were collected at each angle, and the resulting standard deviation for these separate determinations of the relaxation time τ_c was about 5%. Within this uncertainty, no systematic variation of τ_1 with concentration was detected over the limited concentration range studied, as has been predicted by Muthukumar and Freed (24). The variation of τ_c with x is shown in figure 3 for all the samples of both systems, and the resulting variation of τ_i with $\bar{M}_w$ is shown in figure 4.

In the cyclohexane system, the experimental points are represented by the relation

$$\tau_1 = (7.7 \pm 0.3) \times 10^{-8} \, \bar{M}_w^{-(1.42 \pm 0.05)} \mu s,$$

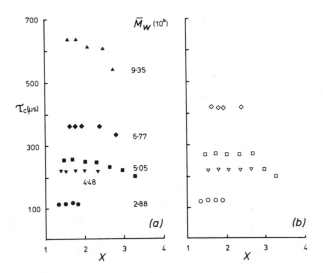

Fig. 3. Variation of τ_c with x for the narrow fractions studied:
(a) cyclohexane at 34.5°C (b) theta mixed solvent.

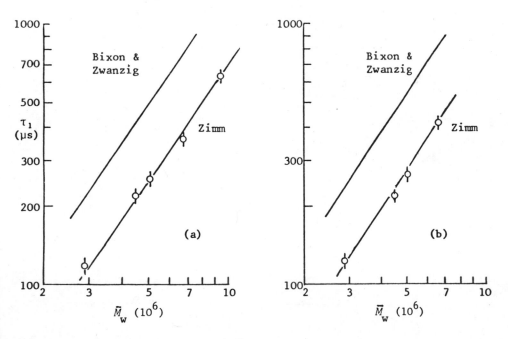

Fig. 4. Variation of τ_1 with $\bar{M}_w$: (a) theta single solvent
(b) theta mixed solvent. Experimental points are
compared with theoretical predictions.

while in the mixed solvent

$$\tau_1 = (7.3 \pm 0.4) \times 10^{-8} \bar{M}_w^{(1.43 \pm 0.05)} \mu s$$

The $\bar{M}_w$ dependence of τ_1 is virtually the same in both systems, and the exponents agree well with the theoretical value of 1.5 for the non-draining model. Furthermore, in both systems, the points are seen to lie on the theoretical line of Zimm, which has been drawn using his value of λ'_1 in equation 1 and a value for $\{\eta\}$ of $88 \times 10^{-3} \bar{M}_w^{\frac{1}{2}}$ ml g^{-1} (23). Thus the results decisively support Zimm's pre-averaged value of 4.04 for λ'_1 rather than Bixon and Zwanzig's non pre-averaged value of 1.96.

The proportion of the scattered intensity due to intramolecular motion, as estimated by the ratio $b/(a + b)$, is shown as a function of x in figure 5 together with the theoretical value expected due to the first mode. The experimental points, taken from a wide range of angular measurements on all the samples, lie slightly above the theoretical line, but perhaps more noteworthy is the fact that they lie within a well defined band, thus indicating that $b/(a + b)$ depends on $\bar{M}_w$ only through the parameter x.

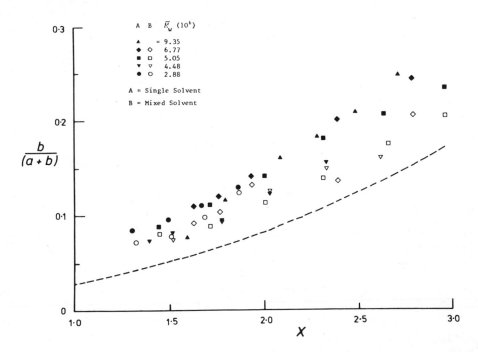

Fig. 5.   Variation with x of the relative intensity of scattering due to intramolecular motion. The line indicated the theoretical prediction.

Initial Slopes

The experimental data have been re-analysed to obtain the initial slopes of the correlation functions so that comparison can be made with the various theoretical predictions. Theoretically, the quantities of interest are the dynamic form factor $P(K,\tau)$ and its normalized initial slope $(-1/T^{**})$ which is used to define an apparent diffusion coefficient D_{app}:

$$\frac{1}{T^{**}} = D_{app} K^2 = \frac{-\lim}{\tau \to 0} \frac{1}{P(K,\tau)} \frac{dP(K,\tau)}{d\tau}$$

$$= \frac{-1}{P(x)} \lim_{\tau \to 0} \frac{dP(K,\tau)}{d\tau}$$

where $P(K,0) = P(x) = (2/x^2)(e^{-x} - 1 + x)$ for a random coil. $P(K,\tau)$ is proportional to $|g^{(1)}(K,\tau)|$, which can be found from the measured normalized intensity correlation function $g^{(2)}(K,\tau)$ since $g^{(2)}(K,\tau) = 1 + f|g^{(1)}(K,\tau)|^2$. (The geometrical factor f is easily determined from a knowledge that $|g^{(1)}(K,0)|$ is unity.) Thus experimentally, $1/T^{**}$ is found from the relation

$$\frac{1}{T^{**}} = \frac{-\lim}{\tau \to 0} \frac{d|g^{(1)}(K,\tau)|}{d\tau}$$

A comparison between experiment and theory is shown in figure 6. In order to display the results from all the samples, a normalized quantity T_R/T^{**} is plotted against x, where T_R is simply $(DK^2)^{-1}$. Büldt's expression (18) for the free-draining case takes a particularly simple form: $(T_R/T^{**})_{fd} = 1 + (P(x)^{-1}$. However, the free-draining case must be ruled out since it predicts $D \propto M$, whereas it is well established that $D \propto M^{\frac{1}{2}}$ for theta systems, in agreement with non-draining theory. For the non-draining case, Freire (17) found

$$(T_R/T^{**})_{nd} = 1 + T_R \frac{R'(x)}{RP(x)}$$

where $R'(x)$ depends on the values of the τ_j's and the elements of the matrix Q. By inserting the factor T_R on the right-hand side of the expression, the explicit dependence of the function on D has been isolated. Freire's curve in figure 6 has been drawn using the theoretical expression for D. However, Freire has subsequently indicated that equation 20 in reference 17 is incorrect and should read:

$$T_R/T^* = P(x) + (3/2^{5/2})R$$

This will make Friere's line in figure 6 coincide with Burchard's prediction (20) which gives better agreement with experimental data.

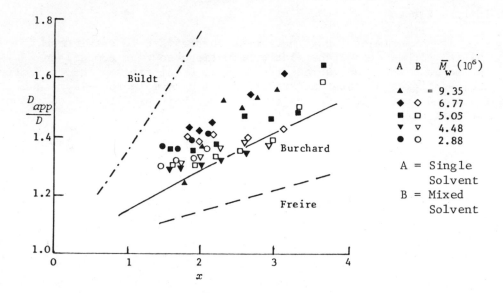

Fig. 6. Variation of D_{app}/D with x taken from the initial
         slopes of the correlation functions for all the samples
         studied. The theoretical predictions are also shown.

     The experimental points cover only relatively low x values,
corresponding to the boundary between the low and intermediate
x-region of Akcasu and Gurol (19). A more searching test of the
theories would require higher x values, in practice greater values
of $<S^2>$, either by using higher $\bar{M}_w$ samples or solutions in good
solvents.

ACKNOWLEDGEMENT

     We should like to thank the Science Research Council for
an equipment grant, and also for the award of a studentship to
one of us (GJ).

REFERENCES

1. W. N. Huang and J. E. Frederick, Macromolecules, 7:34
    (1974).
2. T. A. King, A. Knox and J. D.G. McAdam, Chem. Phys. Lett.,
    19:351 (1973).
3. J. D. G. McAdam and T. A. King, Chem. Phys., 6:109 (1974).
4. T. A. King and M. R. Treadaway, J.C.S. Faraday II, 72:1473
    (1976).
5. R. Pecora, J. Chem. Phys., 40:1604 (1964).
6. R. Pecora, J. Chem. Phys., 43:1562 (1965).

7.   R. Pecora, J. Chem. Phys., 49:1032 (1968).
8.   G. Jones and D. Caroline, Chem. Phys., 37:187 (1979).
9.   G. Jones and D. Caroline, Chem. Phys., 40:153 (1979).
10. S. Lacharjana and D. Caroline, Macromolecules, 10:365 (1977).
11. B. H. Zimm, J. Chem. Phys., 24:269 (1956).
12. J. M. Schurr, Chem. Phys., 30:243 (1978).
13. P. E. Rouse, J. Chem. Phys., 21:1272 (1953).
14. M. Bixon and R. Zwanzig, J. Chem. Phys., 68:1890 (1978).
15. B. E. A. Saleh and J. Hendrix, Chem. Phys., 12:25 (1976)
16. J. Hendrix, B. Saleh, K. Gnädig and L. de Maeyer, Polymer 18:10 (1977).
17. J. J. Freire, Polymer, 19:1441 (1978).
18. G. Büldt, Macromolecules, 9:606 (1976).
19. Z. Akcasu and H. Gurol, J. Polym. Sci. Polym. Phys. Ed., 14:1 (1976).
20. W. Burchard, Macromolecules, 11:455 (1978).
21. Y. Miyaki, Y. Einaga and H. Fujita, Macromolecules, 11:1180 (1978).
22. H. Yamakawa, "Modern Theory of Polymer Solutions", Harper and Row, New York, 1971, p.272.
23. M. Fukuda, M. Fukotomi, Y. Kato and T. Hashimoto, J. Polym. Sci. Polym. Phys. Ed., 12:871 (1974).
24. M. Muthukumar and K. J. Freed, Macromolecules, 11:843 (1978).

QUASIELASTIC LIGHT SCATTERING STUDY OF INTERMICELLAR

INTERACTIONS

Mario Corti and Vittorio Degiorgio*

CISE S.p.a., P.O.B.3986
20100 Milano
Italy

INTRODUCTION

The aim of this paper is to discuss how quasielastic light
scattering measurements can be used to obtain information on
the pair interaction potential between two ionic micelles in
aqueous solution. Although the experimental results reported here
concern only micellar solutions, we believe that the technique
may be equally well applied to any macromolecular or colloidal
solution.

The knowledge of the interaction potential V is useful for
several reasons: i) some micellar parameters can be directly
derived from V, for instance the electrostatic part of V depends
on the electric charge of the micelle; ii) non-ideality effects
can be taken into account, and therefore micellar properties can
be obtained from the experimental results without extrapolating
the experimental data at the critical micelle concentration (cmc);
iii) the phase transitions occurring in micellar solutions may be
better understood.

Micelle formation is a stepwise self-association process
characterized by the fact that the transition from the predominantly
unassociated state to the micellar state does occur over a narrow
critical range of concentration (1-4). For the sake of simplicity
this narrow critical range is characterized by a single concentration
c_o, called the critical micelle concentration (cmc). Solutions with

*Researcher from the Italian National Research Council (CNR).

amphiphile concentrations c larger than c_0 are assumed to
contain a micelle concentration $c-c_0$ and a monomer concentration
c_0. The description of micelle formation in terms of a stepwise
association process implies that micelles are polydisperse and
that the average aggregation number is an increasing function of
concentration c (5).

     Equilibrium and trasport properties of micellar solutions
are generally measured as functions of the amphiphile concentration
c. Individual micelle parameters are obtained by extrapolating
the experimental data at c_0. The concentration dependence of the
solution properties above c_0 is due to the combination of two
effects, intermicellar interactions and change of micelle size
with c. It is generally difficult to separate the two effects.
In the case of ionic micellar solutions the first effect is
certainly the most important at moderate ionic strength whereas
the change of micelle size may be the most significant effect
at high ionic strength.

     The second problem in the interpretation of the experimental
results is the choice of a realistic parametrized expression for
the pair interaction potential of two micelles. This is an
open point which is presently investigated in many theoretical
and experimental groups. We have chosen the interaction potential
of the Derjaguin-Landau-Verwey-Overbeek (DLVO) theory (6). This
potential consists of a hard sphere repulsive part, of an
electrostatic long-range repulsion and of a London-Van der Waals
attraction, and contains only two unknown parameters, the
electric charge of the micelle and the Hamaker constant.

     We report in this paper preliminary measurements of the
average intensity and of the intensity correlation function of
laser light scattered from dilute aqueous solutions of sodium
dodecyl sulfate (SDS) at NaCl concentration in the range
0.1-0.5 M. The experimental data give the apparent molecular
weight M and the diffusion coefficient D of the micellar solution.
The dependence of M and D on the amphiphile concentration c is
found to change considerably with the ionic strength of the
solution. At low ionic strength the interaction potential is
mainly determined by excluded volume and by electrostatic
repulsion. At increasing ionic strength, the Coulomb potential is
screened more and more effectively, the London-Van der Waals
attraction becomes important, and may finally lead to coagulation
of micelles. The electric charge of the micelle and the Hamaker
constant are determined by a fit of the DLVO model to the light
scattering data.

BROWNIAN DIFFUSION OF INTERACTING PARTICLES

We recall in this Section the theoretical results which allow
to connect the concentration dependence of the diffusion
coefficient to the pair interaction potential V, and allow to
give an analytical expression to V.

The classical theory of Brownian motion applies to the
stochastic motion of a large particle due to random collisions
with the much smaller molecules of the surrounding fluid. In
macromolecular solutions the concentration of Brownian particles
can be sufficiently high that the interaction of particles will
affect their diffusion. To the first order in the concentration
c the diffusion coefficient D can be written in the form

$$D = D_o(1 + k_D c)$$

where c (g/cm^3) is the mass of solute particles per unit volume of
solution. Theoretical results are usually expressed in terms of
the volume fraction ϕ of the particles, that is

$$D = D_o(1 + k_D'\phi)$$

where $k_D' = k_D/\bar{v}$, and $\bar{v}$ is the specific volume of the solute
particles. The calculation of k_D' considers two types of interactions, the first due to inter-particle forces and the second
resulting from the fact that movement of one particle through
the fluid generates a velocity field which affects the motion of
neighbouring particles. The general theoretical expression
for k_D' can be written as

$$k_D' = k_I' - k_f', \tag{1}$$

where k_I' represents the static contribution and is therefore
proportional to the second virial coefficient, and k_f' is the
dynamic part which takes into account the volume fraction
dependence of the friction coefficient. Eq.(1) is derived from
the generalized Einstein relation (7)

$$D(\phi) = \frac{\phi}{1-\phi} \frac{(\delta\mu/\delta c)_{p,T}}{f(\phi)} \tag{2}$$

where μ is the chemical potential and f is the friction coefficient
of solute particles.

Considering the case of rigid spherical particles of radius
a, with a pair interaction potential V(x), where x = (R-2a)/2a
and R is the distance between the centre of the two particles,
k_I' is expressed by the relation

$$k'_I = 8 + 24 \int_0^\infty dx\ (1+x)^2 \left[1-e^{-V(x)/k_BT}\right] \tag{3}$$

where 8 is the hard sphere contribution. Whereas the expression
for k'_I is well-established, there is still some discrepancy
among the calculations fo k'_f presented by different authors.
It is possible to write down a relation for k'_f which has the same
structure as Eq.(3), that is

$$k'_f = k'_{fo} + \int_0^\infty dx\ F(x) \left[1-e^{-V(x)/k_BT}\right] \tag{4}$$

where k'_{fo} is the hard-sphere contribution.

The most complete treatments appear to be those by Batchelor
(7) who obtains

$$k'_{fo} = 6.55, \text{ and}$$

$$F(x) = 11.89\ (1+x) + 0.706-1.69\ (1+x)^{-1} \tag{5}$$

and by Felderhof (9) who finds

$$k'_{fo} = 6.44, \text{ and}$$

$$F(x) = 12(1+x) - \frac{15}{8}\ (1+x)^{-2} + \frac{27}{64}\ (1+x)^{-4} + \frac{75}{64}\ (1+x)^{-5} \tag{6}$$

An earlier treatment by Pyun and Fixman (10), reformulated
by Goldstein and Zimm (6), gives

$$k'_{fo} = 7.16, \text{ and}$$

$$F(x) = 12(1+x) + \frac{15}{8}\ (1+x)^{-2} - \frac{3}{4}\ (1+x)^{-4} + \frac{15}{128}\ (1+x)^{-6} \tag{7}$$

A critical comparison of all the published approaches can be
found in Felderhof's paper.

The interaction potential $V(x)$, as it is usually written
in colloid physical chemistry (6), is the sum of a repulsive
interaction V_R and an attractive London-Van der Waals interaction
V_A. The expression of V_A derived by Hamaker (11) for the case of
two spheres is

$$V_A = -\frac{A}{12} \left[(x^2+2x)^{-1}+(x^2+2x+1)^{-1}+2\ln(x^2+2x)/(x^2+2x+1)\right] \tag{8}$$

where A is the Hamaker constant. Eq. (8) does not take into account
retardation effects and therefore is not valid for large values
of x (6). The range of ionic strengths considered in this paper is
such, however, that the region of large separations between the
particles is not relevant for the fit to the experimental results.

The repulsive interaction is due to the electric charge of
the spheres. The layer of ions surrounding the sphere is
divided into two regions a thin inner region called the Stern layer
and a diffuse outer region called the Gouy layer where the ions
are treated as point charges obeying the Poisson-Boltzmann
equation. In this picture the long-range repulsion is determined
by the potential of the diffuse layer Ψ_o. Unfortunately no exact
analytical solution of the Poisson-Boltzmann equation is available.
Approximate solutions may be derived by assuming that $\Psi_o < k_B T/e$,
where e is the electronic charge. Under this assumption, the
Poisson-Boltzmann equation reduces to

$$\nabla^2 \Psi = \kappa^2 \Psi \tag{9}$$

where κ is the Debye-Hückel reciprocal length parameter given
by

$$\kappa^2 = 8 \pi c_s e^2 z^2 / \varepsilon k_B T, \tag{10}$$

where ε is the dielectric constant of the suspending medium, z
is the valence of the ionic species in solution, and c_s is the
concentration (ions/cm^3) of the same species. Approximate
expressions for the interaction potential V_R have been derived
from Eq. (9) in the two limit cases of $\kappa a \ll 1$ and $\kappa a \gg 1$. In the
former case, the expression of V_R is (12)

$$V_R = \frac{q^2 e^2}{2\varepsilon a (1+\kappa a)^2} \frac{e^{-2\kappa a x}}{(1+x)} \tag{11}$$

where qe is the electric charge of the particle.

The latter case gives (13,14)

$$V_R = (\varepsilon a \Psi_o^2/2) \ln \left[1 + \exp(-2\kappa a x)\right] \tag{12}$$

where the surface potential Ψ_o is related to the charge qe
through the expression

$$\Psi_o = (2k_B T/e) \sinh^{-1} \left[2 \pi e \kappa^{-1} q e/(4 \pi a^2 \varepsilon k_B T)\right] \tag{13}$$

We show in Figure 1 the behaviour of $V(x)/k_BT$ for a set of salt concentrations. In order to perform the numerical computation it is useful to write Eq.(12) as follows

$$\frac{V_R(x)}{k_BT} = \frac{\varepsilon \, k_BT}{2 \, e^2} \, a \left(\frac{e\Psi_o}{k_BT}\right)^2 \, \ln \left[1 + \exp \, (-2\kappa ax) \right] \qquad (12')$$

The plots have been computed by using Eqs. (8) and (12') with the following values for the parameters: $a = 25$ Å, $T = 298°K$, $A = 13.8 \, k_BT$, $e\Psi_o = 2.5 \, k_BT$.

The expression of k_I becomes particularly simple if $\kappa a \ll 1$, $V_R/k_BT \ll 1$, and the contribution of V_A can be neglected. Indeed, by putting $1-\exp \, \{-V(x)/k_BT\} \approx V(x)/k_BT$ into Eq. (3) and using for $V(x)$ the expression given in Eq.(11), it is easy to derive that

$$k_I = 8 + \frac{q^2}{2\bar{v} \, M \, m_s}$$

where M is the molecular weight of the particles, and m_s (mole/cm^3) is the electrolyte concentration. This is the formula used in

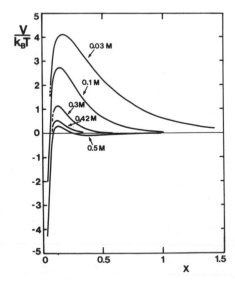

Fig. 1. Plots of the pair interaction potential $V(x)$ as a function of the reduced and normalized distance x at five distinct salt concentrations. Values of the parameters are given in the text.

several papers (15) to compute the micellar charge from the experimentally determined k_I.

It is important to realize that the Hamaker expression for V_A makes the integrals which appear in Eq.(3) and (4) divergent. To eliminate the divergence a lower cut-off $x_L>0$ has to be imposed. The physical origin of x_L is that the two particles cannot approach closer than the Stern layer thickness (6).

It is easy to see from the structure of Eq.(3) that a purely repulsive interaction potential produces a positive k_I, and that k_I is larger when the amplitude and (or) the range of the repulsive potential are larger. The attractive part of the potential tends to reduce k_I, and may lead, if large enough, to negative values of k_I. When $k_I<0$, attraction is more effective than repulsion, and consequently the particles in solution may aggregate. Of course, a thorough discussion of the particle aggregation process cannot be made by considering only k_I which contains an integral information of $V(x)$, but the shape of the interaction potential should also be taken into account.

Particularly important for the stability of the colloidal solution are the maximum value V_{max} and the minimum value V_{min} = $V(x_L)$ (see Figure 1). The positive potential barrier V_{max} will be typically a few k_BT's at low ionic strength and even smaller at large ionic strength. The negative V_{min} which represents the binding energy between two particles may be so large in comparison with k_BT that the aggregation becomes practically irreversible. In this case the colloidal solution will not be stable because particles coming from large separations will always acquire, sooner or later, the kinetic energy necessary to overcome the potential barrier, whereas aggregated particles will never be able to split. On the contrary, if the absolute value of the binding energy V_{min} between the two particles is not large in comparison with k_BT, aggregation is reversible and it will be possible to observe by increasing the ionic strength of the solution stable solutions of polydisperse aggregates. In any case, however, by further increasing the ionic strength precipitation (the so-called salting out) will finally occur.

The discussion made above about the sign of k_I can be equally well applied to k_D. This consideration may be easily justified by considering that the leading term in $F(x)$ (see Eq.(4)) is the first one, 12 (1+x). Note incidentally that the leading term coincides in the formulations by Felderhof and by Pyun and Fixman, and is only slightly different in that by Batchelor. Since the factor 1-exp $(-V(x)/k_BT)$ is weighted in the integral appearing in the expression of k_D by a weight distinct from that appearing in Eq. (3), the condition $k_I=0$ may not coincide with the condition $k_D=0$.

EXPERIMENTAL RESULTS

The quasielastic light scattering apparatus and the preparation procedure of the materials used for the present investigation have been already described in previous papers (16,17). The alkyl chain composition of sodium dodecyl sulfate (SDS) was checked by gasliquid-chromatography on fatty alcohol recovered after acid hydrolytic clearage, and resulted as a 99,7% C_{12} pure fraction.

Intensity correlation functions were measured at 25°C on NaCl aqueous solutions of SDS in the NaCl concentration range 0.1-0.5 M and in the SDS concentration range 5-30 mg/cm^3. The mass diffusion coefficient D is derived in our experiment by fitting the time dependent part of the measured correlation function to a single exponential. The decay time of the exponential, τ_c, is connected to D by the relation $\tau_c=(2k^2D)^{-1}$, where k=$(4\pi n/\lambda)$ sin $\theta/2$, n is the index of refraction of the solution, λ the wavelength of laser light, and θ the scattering angle. For our experiment, λ = 5145 Å and θ = 90°. Since micelle polydispersity would produce deviations from single exponential behaviour, we have applied systematically to our data the usual cumulant fit (18,19) to obtain also the fractional variance v = $<\delta D^2>/<D>^2$.

Some examples of experimentally obtained correlation functions are shown in Figure 2. The obtained diffusion coefficients are plotted in Figure 3 as a function of the amphiphile concentration for six distinct salt concentrations.

The experimental data have been fitted with the linear function

$$D = D_o \left[1 + k_D (c-c_o) \right] \tag{15}$$

The obtained values of D_o and k_D, together with the values of c_o, derived from Refs. (15)and (20),are reported in Table 1.

The micelle diffusion coefficient D_o decreases by increasing the salt concentration m_s. This effect is expected because it is known from average scattered intensity measurements that the micellar size increases with m_s. The slope k_D is a decreasing function of m_s which becomes equal to zero at m_s = 0.4 M, and negative at larger values of m_s. The change in the sign of k_D going from low to high ionic strength is similar to the change in the sign of k_I reported by Emerson and Holtzer (15), by Anacker et al., (21), and also observed by us (19). We recall that k_I is derived from scattered light intensity data (19) by plotting the reciprocal of the apparent molecular weight M as a function of the reduced amphiphile concentration, that is, $M^{-1} = M_o^{-1}\{1+k_I(c-c_o)\}$, where M_o is the molecular weight of the SDS micelle at the critical

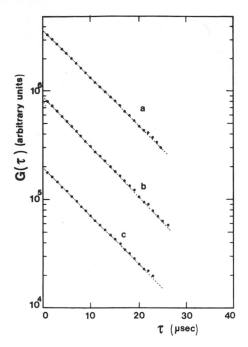

Fig. 2.   The time-dependent part $G(\tau)$ of the intensity-correlation
          function of SDS solutions at 25°C and 0.2 M NaCl. Curve
          a) refers to an amphiphile concentration of 20 mg/cm^3,
          b) 10 mg/cm^3, c) 5 mg/cm^3.

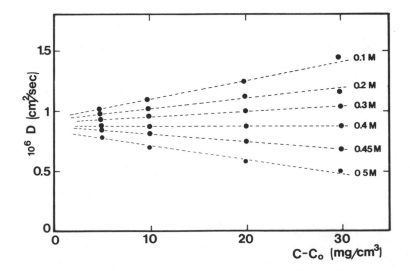

Fig. 3.   The diffusion coefficient D plotted as a function of the
          reduced amphiphile concentration c-c$_o$ at six distinct salt
          concentrations.

TABLE 1

The critical micelle concentration c_o, the extrapolated diffusion coefficient D_o, the hydrodynamic radius R_H, the fractional variance v, the experimental and theoretical slopes k_D reported as functions of NaCl molar concentration m_s.

The values of c_o are taken from Refs.(15) and (20). The theoretical k_D is computed by using the best fit parameters of Eq. (20).

m_s (M)	c_o (mg/cm^3)	$10^6 D_o$ (cm^2/s)	R_H (Å)	v	k_D (cm^3/g) exp.	k_D (cm^3/g) theor.
0.10	0.43	0.96 ± 0.02	25.3	0.05 ± 0.03	14.4 ± 1.5	14.5
0.20	0.27	0.94 ± 0.02	25.6	0.04 ± 0.03	9.3 ± 1.5	6.0
0.30	0.24	0.91 ± 0.02	26.2	0.03 ± 0.03	5.1 ± 1.5	3.4
0.40	0.17	0.87 ± 0.02	27.1	0.05 ± 0.03	0 ± 1.5	0
0.45	0.16	0.87 ± 0.02	27.0	0.09 ± 0.03	-6.7 ± 1.5	-3.1
0.50	0.15	0.84 ± 0.02	27.9	0.17 ± 0.03	-13.9 ± 1.5	-7.2

micelle concentration. The salt concentration at which $k_I = 0$ is $m_s = 0.42$ M NaCl. The fractional variance v obtained through the cumulant fit from the experimental correlation functions measured at c=20 mg/cm^3 is also reported in Table 1. At low ionic strength, v is zero within the experimental errors, whereas at large ionic strength, when k_D becomes negative, v markedly increases.

Light scattering measurements on SDS solutions at high ionic strength (0.6 M Nacl) have been reported by Mazer et al (22,23), and by ourselves (19,24). We simply recall here the most relevant features of these experimental results: i) both D and M^{-1} are decreasing functions of $c-c_o$, it is however difficult to derive from the data the slopes k_D and k_I because D and M^{-1} are nonlinear functions of the reduced concentration in the interval $c_o < c < 30$ mg/cm^3. Since D and M^{-1} decrease considerably in a very small concentration interval above c_o, values of k_D and k_I obtained by extrapolating the data relative to concentrations much larger than c_O are not correct; ii) the shape of the intensity correlation function reveals that the solution is strongly poly-disperse; iii) micellar parameters are considerably temperature dependent; iv) the aggregates are not spherical.

DISCUSSION

The hydrodynamic radius R_H of the micelle at the cmc is derived from D_O by using the Einstein-Stokes relation $R_H = k_B T/(6 \pi \eta D_O)$, where the viscosity of the solution η is also dependent on m_s. The values of R_H reported in Table 1 are much larger than the length of the fully stretched SDS monomer ℓ which is 20-21 Å. It should be considered that the actual length of the monomer inside the micelle may be smaller than ℓ because, just as found for lipid bylayers, a strongly increased disorder toward the chain ends is likely to exist in the micelle core (1). It seems therefore necessary to attribute a non-spherical form to SDS micelles near the cmc in the investigated range of salt concentra-tions. For a more quantitative discussion of micellar shape we refer to our previous works (17,19). Recently Aniansson (25) presented in a very interesting paper a dynamic model of the micellar surface with "frequent and rather large protrusions of monomers out from the hydrophobic core". The protrusion of monomers would lead to an increased effective size of the micelle. It appears however difficult to justify by means of this dynamic model the large difference between the experimental R_H and the calculated actual length of the monomer.

The dependence of D on the amphiphile concentration may be generally due to two factors, the first of which is the effect of intermicellar interactions and the second is the possible change of micelle size with concentration. To put the discussion on a quantitative basis we will assume a concentration dependence of the weight-average micellar molecular weight of the type

$$M = M_o \left[1 + k_M (c - c_o) \right] \qquad (16)$$

where k_M is >0. It can be shown that the effect of Eq.(16) on k_D is that of subtracting to the concentration slope calculated purely on the basis of interactions a term $k_M/3$.

By extending to micellar systems some general methods introduced to describe multiple equilibria in stepwise self-association processes, Mukerjee (5) has shown that the concentration dependence of the number-average molecular weight M_n can be derived from the concentration dependence of the weight-average molecular weight as follows

$$\frac{c-c_o}{M_n(c)} = \int_o^c \frac{d(c-c_o)}{M(c)} \tag{17}$$

By using Eq.(16) and the approximation $k_m(c-c_o) \ll 1$, Eq.(17) gives

$$M_n = M_o \left[1 + \frac{k_M}{2} (c-c_o) \right] \tag{18}$$

Therefore the variance v_M of the molecular weight probability distribution, defined as $v_M = M/M_n - 1$, is given by

$$v_M = \frac{k_M}{2} (c-c_o) \tag{19}$$

Eq.19 is interesting because it establishes a very simple and direct connection between polydispersity and change of aggregation number with concentration. Let us note, incidentally, that the parameter k_m defined in Ref. (17) refers to the concentration dependence of M_n and is therefore equal to $k_M/2$.

In order to obtain a rough evaluation of the interaction potential by using our preliminary data, we have made the simplifying assumption that $k_M = 0$, so that k_D and k_I can be computed by considering only Eqs. (3) and (4). We have used the results obtained by Felderhof and expressed by Eq. (6). A numerical test of the effect produced on k_D by using the function $F(x)$ given by Eqs.(5) or (7) instead of Eq.(6) was also performed. These effects are completely negligible in all practical cases. The interaction potential $V=V_R+V_A$ is given by Eqs.(8) and (12'), where a is taken to coincide with R_H. The unknown parameters are the electric potential Ψ_o and the Hamaker constant A. The two experimental conditions:

$$k_D = 0 \quad \text{at} \quad 0.4 \text{ M NaCl}$$

$$k_I = 0 \quad \text{at} \quad 0.42 \text{ M NaCl}$$

allow to determine through a best fit procedure the two unknown parameters. The result is

$$e\Psi_o = 2.5 \ k_B T \simeq 0.06 \text{ eV}$$
$$A = 13.5 \ k_B T \simeq 5 \times 10^{-20} \text{ J} \tag{20}$$

By using the numerical values given by Eq.(20) we have computed the slope k_D for different salt concentrations. The results are given in Table 1. The agreement with the experimental data is reasonable, and seems to indicate that Ψ_o and A do not depend on the ionic strangth of the solution, or, at least, their dependence is weak.

As a conclusion, we list below several considerations and comments:

a) the micellar charge number q can be derived from Eqs.(15) and (20). We obtain q $\simeq$ 30. This is the charge of the micelle including the Stern layer.

b) Eqs.(8) and (12) apply only to spherical particles. Micelles, in the investigated range of salt concentrations are not spherical, but they can still be considered as globular particles.

c) One could try to evaluate k_M from polydispersity data. By taking, for instance, $V_M=0.03$ at c=30 mg/cm^3, one finds from Eq.(19) $k_M=2$. This is probably the correct order of magnitude of k_M when m_s is smaller than 0.4 M NaCl.

d) The fact that k_M is small but not equal to zero may have some connection with the experimental finding that k_I becomes zero at a salt concentration distinct from that which makes k_D equal to zero.

e) The best fit computations have been performed with a lower cut-off $x_L=4\times10^{-2}$ which corresponds to a distance of closest approach of 1 Å between the two micelles. We have verified that the best fit parameters appearing in Eq.(20) are not considerably modified if x_L is chosen between 4 and 8×10^{-2}.

f) Unpublished results of A.Lips, (26) also analyzed with the DLVO theory, gave best fit parameters similar to those we report here.

g) The obtained value for Ψ_o is in contradiction with the assumption $e\Psi_o<<k_BT$ introduced to linearize the Poisson-Boltzmann equation. Numerical studies of the Poisson-Boltzmann equation indicate however that the linear approximation is not too bad even when $e\Psi_o \sim 2k_BT$ (14).

We acknowledge the help of P.Zappa for the purification of SDS samples. This work was supported by CNR/CISE contract no. 7800901.02.

REFERENCES

1. C. Tanford "The Hydrophobic Effect", Wiley, New York (1973)
    J. Phys. Chem., 78:2469 (1974).
2. R. Nagarajan and E. Ruckenstein, J.Colloid Interface Sci.,
    60:221 (1977).
3. J. N. Isrealashvili, D. J. Mitchell and B. W. Ninham,
    J.Chem.Soc. Faraday Trans.,II:72, 1525 (1976).
4. A. Wulf, J. Phys. Chem., 82:804 (1978).
5. P. Mukerjee, J. Phys. Chem., 76:565 (1972).
6. B. Goldstein and B. H. Zimm, J. Chem. Phys., 54:4408 (1971).
7. G. K. Batchelor, J. Fluid Mech., 74:1 (1976).
8. P. Doty and J. T. Edsall,Advances in Protein Chemistry,
    6:35 (1951).
9. B. U. Felderhof, J. Phys. A., 11:929 (1978).
10. C. W. Pyun and M. Fixman, J. Chem. Phys., 41:937 (1964).
11. H. C. Hamaker, Physics, 4:1058 (1932).
12. D. Stigter and T. L. Hill, J. Phys. Chem., 63:551 (1959).
13. B. Derjaguin, Kolloid-Z, 69:155 (1934), Acta Physicochim,
    10:333 (1939).
14. R. Hogg, T. W. Healy and D. W. Fuerstenau, Trans. Faraday Soc.,
    62:1638 (1966).
15. M. F. Emerson and H. Holtzer, J. Phys. Chem., 71:1898 (1967).
16. M. Corti and V. Degiorgio, Chem. Phys. Letters, 49:141 (1977).
17. M. Corti and V. Degiorgio, Chem. Phys. Letters, 53:232 (1978).
18. D. E. Koppel, J. Chem. Phys., 57:484 (1972).
19. M. Corti and V. Degiorgio, Ann. Phys., (Paris), 3:303 (1978).
20. H. F. Huisman, Proc. Koninkl. Ned. Akad. Wetenschap., 67B:367,
    376, 388, 407 (1964).
21. E. W. Anacker, R. M. Rush, and J. S. Johnson, J. Phys. Chem.,
    76:565 (1972).
22. N. A. Mazer, G. B. Benedek and M. C. Carey, J. Phys. Chem.,
    80:1075 (1976).
23. C. Y. Young, P. J. Missel, N. A. Mazer and G. B. Benedek,
    J. Phys. Chem., 82:1375 (1978).
24. M. Corti and V. Degiorgio, in "Solution Chemistry of Surfactants",
    K. L. Mittal ed., Plenum, New York (1979), p.377.
25. G. E. A. Aniansson, J. Phys. Chem., 82:2805 (1978).
26. A. Lips, private communication.

STABILITY AND FUSION OF VESICLES

Nicole Ostrowsky and Didier Sornette

Laboratoire de Physique de la Matière Condensée
associé au CNRS (LA 190)
Université de Nice, Parc Valrose. 06034
Nice Cedex France

Amphiphilic molecules, when put into water solutions, build up organized structures so as to shield their hydrophobic parts from the water molecules.

The purpose of this paper is to review the theoretical explanation for the formation of such structures (§I) and then show (§II) how light scattering can be useful in measuring the size and the stability of such structures. We will concentrate mostly on the spherical bilayer structure, called a vesicle, which is being extensively studied as an idealized membrane model. In particular, the possibility of fusion between different vesicles is of great interest for biological purposes. We will present at the end of the paper a simple pore model which can help understand the evolution in size of the vesicles we have studied.

I.   A.   FORMATION OF ORGANIZED STRUCTURES

The theory of amphiphilic molecules assemblies in water solution has been extensively reviewed by different authors (1,2). It basically involves three steps:

a) Statistical thermodynamics yields the molar fraction x_N of amphiphilic molecules participating in a structure involving N amphiphilic molecules, the so-called "N-state", as a function of the monomer fraction x_1, and the lowering in free energy $(\mu_1-\mu_N)$ which occurs when one molecule goes from the monomer state to the N- state:

$$X_N = NX_1^N \exp\left[\frac{N(\mu_1 - \mu_N)}{kT}\right] \tag{1}$$

X_N is therefore the product of two terms : the entropy term NX_1^N which favours small structures, and the free energy term

$$\exp\left[\frac{N(\mu_1 - \mu_n)}{kT}\right]$$

which now remains to be evaluated.

b) Simple model for the free energy difference $(\mu_1 - \mu_N)$: This difference has two major contributions, the bulk part and the surface part.

The bulk part arises both from:

- the difference in self energy of the hydrocarbon tails in water and in bulk hydrocarbon.

- the difference in electrostatic energy of the polar head in the monomer state and in the N-state. It must be noted that this bulk part is N- independent and can be included in some normalization factor.

The surface part of $(\mu_1 - \mu_N)$ actually comes from the μ_N term. It can be derived, in a simplistic model, in terms of the free surface a per polar head in the aggregated N- state, and results from two opposing phenomena.

- The hydrophobic effect, characterized by an interfacial energy proportional to a and which tends to bring the amphiphilic molecules closer to one another.

- The repulsion between neighbouring amphiphilic molecules. This repulsion comes mainly from the polar heads which are either charged, or zwitterionic. In a first approximation, this repulsion can be accounted for by a term inversely proportional to a.

Introducing the phenomenological parameters γ and C, the surface part of μ_N can be written as:

$$(\mu_N)_s = \gamma a + \frac{C}{a} \tag{2}$$

γ, the interfacial free energy per unit area has been experimentally measured and is typically on the order of 50 erg/cm^2.

As for C, it can be computed for zwitterionic polar heads of dimension D large compared to $\sqrt{a}$. It is simply derived from the energy of a planar capacitor:

$$C \approx e^2 D/2\epsilon_o \epsilon_r$$

where e is the electronic charge, and ϵ_r, at the water-hydrocarbon interface, is roughly half its value in pure water. Minimizing $(\mu_N)_s$ with respect to a leads to an optimal surface area $a_o = \sqrt{C/\gamma}$. This area depends on the nature of the polar heads, but also on the physical state of the aliphatic chains which are rigid below their transition temperature T_ϕ and become flexible above T_ϕ. This experimentally observed fact (3) possibly reflects a different packing of the polar heads below and above T_ϕ, as well as a difference in the Van Der Waals attraction energy of "kinked" chains compared to rigid (all trans) chains (4). Figure 1 roughly shematizes the different configurations of the bilayer below and above the phase transition of the chains.

Rewriting equation (2) as a function of the optimal surface area a_o leads to:

$$(\mu_N)_s = 2\gamma a_o + \frac{\gamma}{a}(a - a_o)^2 \qquad\qquad (3)$$

However, the minimum value $2\gamma a_o$ of $(\mu_N)_s$ may not always be reached, due to the packing constraints existing in an N- state, and which are summarized below.

c) <u>Packing constraints</u>: The three geometrical parameters a, v, and ℓ relevant to describe the chains in an N- state are represented on the left part of figure 2.

The free surface a per polar head has been studied in b). The concept of a volume v occupied by the chains amounts to approximating the repulsive interaction term between chains by a hard core potential. It has been experimentally measured and can be expressed in terms of the number n of carbon atoms in the chain (1):

$$v/chain \approx (27.4 + 26.9n) \ \text{Å}^3$$

As for the length ℓ of the chains, it cannot exceed a certain critical length ℓ_c which, in the liquid state, is taken generally as 80% of the chain length in its most extended configuration:

$$\ell < \ell_c \quad \text{with (1)} \quad \ell_c \approx (1.5 + 1.27n) \times 0.8 \ \text{Å}$$

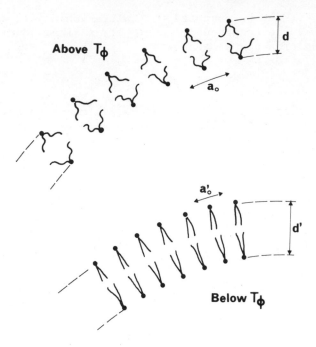

Fig. 1.   Schematized configuration of a bilayer above and below
           the aliphatic chains phase transition.

           Numerical values for DPL are: $a_0 \simeq 58 \ \overset{o}{A}{}^2$, $d \simeq 37 \ \overset{o}{A}$,
           $a_0^1 \simeq 48 \ \overset{o}{A}{}^2$, $d' \simeq 47 \ \overset{o}{A}$.

The packing constraints can be illustrated on a few simple
examples: - spherical micelle (fig. 2a):

The micelle of radius ℓ must accomodate the N chains in its
volume $4/3 \ \pi\ell^3$ and the N polar heads on its surface $4\pi\ell^2$ which
yields the relation

$$v/a \ = \ \ell/3$$

As ℓ cannot exceed ℓ_c, the packing of amphiphiles into spherical
micelles will only be possible if

$$v/a \leqslant \ell_c/3$$

- cylindrical micelles of radius ℓ (fig. 2b): Following the
same line of reasoning as above, the condition for packing the
amphiphiles in such a structure is given:

$$v/a \leqslant \ell_c/2$$

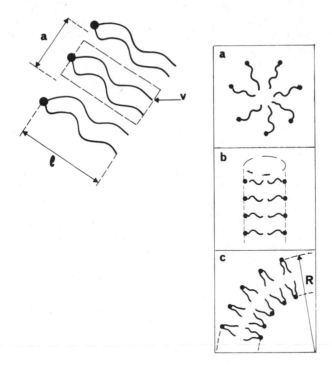

Fig. 2.   Illustration of the packing constraints for a spherical
          micelle (a) a cylindrical micelle (b) and a spherical
          vesicle (c).

- spherical vesicles (fig. 2c) of external radius R: the
geometrical constraints for the outerlayer molecules lead to
the following relation between v, a and ℓ:

$$\frac{v}{a} = \ell(1 - \frac{\ell}{R} + \frac{\ell^2}{3R^2})$$
(4)

which defines a critical radius R_c above which the external
polar heads may occupy their optimum free surface a_o:

$$R_c \simeq \ell_c / (1 - v/a\ell_c)$$
(5)

Writing up the equivalent packing constraints relation for the
internal layer leads to a negative critical radius. This amounts
to saying that there are no packing constraints for the internal
layer: the number of amphiphilic molecules can always be adjusted
so that the polar heads occupy their optimum surface area and
the chains fill up their alloted volume.

Numerical data for n = 16 (palmitic chain) and $a_o \simeq 60$ Å^2 (phosphatidylcholine polar head's free surface in the liquid state) yield the ratio v/a_o u 7.7 Å per chain, and ℓ_c = 17.4 Å. Single chain amphiphiles (lysolecithin) for which

$$\ell_c/3 < v/a < \ell_c/2$$

can therefore assemble into cylindrical micelles. As for the double chain dipalmitoyl phosphatidylcholine (DPL) we have studied, $v/a_o \simeq 15.4$ Å which prohibits a micellar structure. Vesicles however can be formed, with a critical radius $R_c \simeq 150$ A. Such small vesicles have indeed been observed (5), provided sufficient energy (sonication) is available in the preparation procedure: although the smaller vesicles are thermodynamically favoured, their formation goes through high energy configurations and they do not form "spontaneously".

We will now briefly discuss the vesicle's size distribution (§ I.B.a) and moreover include the important curvature effect (§ I.B.b) we have neglected until now.

I.   B.   SIZE DISTRIBUTION OF VESICLES

a) the simple model for the free energy $(\mu_N)_s$ given above (equ. 3) neglects the curvature effect which will be considered later. Taking into account the geometrical constraints yields the following:

- for $R \geqslant R_c$ $(\mu_N)_s = 2\gamma a_o$ (6)

- for $R < R_c$ $(\mu_N)_s = 2\gamma a_o$ for the inner layer

$(\mu_N)_s = 2\gamma a_o + \frac{\gamma}{a}(a-a_o)^2$ for the outer layer

Computing the numbers of molecules on the outer and inner layers, and using equation (4) to express a in terms of R, one can derive for $(\mu_N)_s$ an average value:

$$\overline{(\mu_N)_s} = 2\gamma a_o + \frac{4\pi\ell_c^2\gamma}{N}\left(1 - \frac{R}{R_c}\right)^2$$ (7)

Making use of the statistical distribution (1), one can express the mole fraction distribution X_N of amphiphiles in N-States, as a function of X_{N_c} :

$$X_{N_C} = N_c (X_1)^{N_C} \exp - \left[N_c \frac{2\gamma a_0}{kT} \right]$$

where N_C is the number of amphiphilic molecules in a vesicle of radius R_C. One obtains the following distribution:

$$X_{N>N_C} = N \left(X_{N_c} / N_c \right)^{N/N_C}$$

$$X_{N<N_C} = N \left(X_{N_c} / N_c \right)^{N/N_C} \exp \left[\frac{-4\pi \ell_c^2 \gamma}{kT} \left(1 - \frac{R}{R_c} \right)^2 \right]$$

which can be approximated by a Gaussian distribution of width:

$$\Delta \simeq (R_c/\ell_c) \sqrt{kT/8\pi\gamma}$$

For DPL, we find $R_C \simeq 150 \overset{o}{A}$ and $\Delta \simeq 5 \overset{o}{A}$.

Before confronting these theoretical predictions with experimental data, we must include the curvature effects particularly important for the small vesicles we have studied.

   b) Curvature effects: they amount to reducing the repulsion term C/a fo eq. (2) since polar heads of finite size D assembled on a spherical surface will have a weaker electrostatic inter-action energy than an equally charged assembly on a planar surface. For zwitterionic polar heads, the corrected repulsion term can be evaluated, using the concept of a spherical capacitor. This leads to the following surface term for μ_N:

$$(\mu_N)_s = \gamma a + \frac{C}{a(1+D/R)}$$

the + (-) sign corresponding to a molecule on the outer (inner) layer of the vescle. On the average, this leads, to first order in D/R:

$$(\mu_N)_s \simeq \gamma a + C/a - 4\pi D\gamma d/N$$

where d is the thickness of the vesicle's wall, The curvature effect can thus be accounted for by an additional negative term, inversely proportional to N, which will shift the vesicle's size distribution towards smaller structures.

   The maximum will correspond to:

$$R_{peak} \simeq R_c (1 - d\, D/\ell_c^2) \text{ with } d \simeq 2\ell$$

and the distribution will be somewhat broadened.

For the DPL vesicles, the size distribution should peak around 80 Å with a total width of 15 Å ($D \simeq 6$ Å for phosphatidylcholine polar heads).

II   A.   EXPERIMENTAL RESULTS

The experiments were carried out in collaboration with C. Hesse-Bezot and P. Bezot.

DPL was purchased from Koch light laboratories and sample purity was checked by thin layer chromatography.

Evaporating a chloroform solution, the lipids were deposited regularly on the walls of a vessel, then mixed on a vortex with a buffer solution (0.1M NaCl, 0.01M Tris, pH = 7.5, 0.01% in weight sodium azide to retard bacterial growth). The sonication was performed above the phase transition temperature, and the samples ultracentrifuged twice (20 mn at 100,000 g and 2 hours at 140,000 g) at 15°C.

The experimental set up is shown on figure 3 and allows both homodyne and heterodyne detection. This latter detection mode proved to be useful in case of polydisperse samples as it has two main advantages.

The background is easily fit by the computer as a linear parameter, and it avoids stray heterodyne components which may distort a homodyne spectrum.

The data were fit to a first order correlation function for the scattered electric field of the form

$$g^{(1)}(\tau) = \int_{0}^{\infty} G(\Gamma) \exp(-\Gamma\tau)d\Gamma$$

with $G(\Gamma) = A_o \delta(\Gamma-\Gamma_o)$ simple exponential fit

$\qquad G(\Gamma) = A_1 \delta(\Gamma-\Gamma_1) + A_2 \delta(\Gamma-\Gamma_2)$ bimodal fit

$$G(\Gamma) = A \exp \left[-\frac{(\Gamma-\bar{\Gamma})^2}{2\bar{\Gamma}^2 v} \right]$$ Gaussian fit (6)

Samples studies at room temperature within a few hours of the
second centrifugation seemed to be fairly monodisperse as a
bimodal fit showed that 90% of the light was scattered from
vesicles having a radius R_1 = 110 Å, and 10% from larger vesicles
with radius R_2 = 340 Å. As the scattered intensity (for a
scattering vector q) from vesicles of external radius R and
internal radius R' grows like (see upper right corner of figure 4):

$$I(R) \propto (sinqR - qR \cos qR - sinqR' + qR' \cos qR')^2$$

our results showed that more than 99% of the vesicles were of the
small kind.

Neglecting the small contribution from the larger vesicles,
we studied the size distribution of the smaller ones with a
Gaussian fit whose results are shown on figure 4. This figure
schematizes the basic steps for going from the $G(\Gamma)$ distribution
to the vesicle's size distribution N (R) and the results are
constent with the theory presented in § I.

However, in a great number of experiments (see for example
fig.5) we have shown that these vesicles are not stable and that
they will tend towards a broader equilibrium distribution shifted
towards the larger sizes. This equilibrium will be reached very
much faster for temperatures closer to the phase transition
temperature T_ϕ of the aliphatic chains. This fact strongly
supports the interpretation of this phenomenon in terms of fusion
between vesicles, as opposed to aggregation which should not have

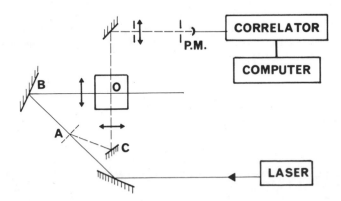

Fig. 3.   Experimental set up: the beam splitter A provides the
          heterodyne reference beam whose intensity and (or)
          polarization may be adjusted. The optical paths ABO and
          ACO are made equal.

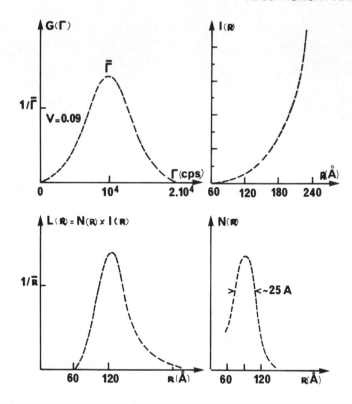

Fig. 4.   Basic steps for going from the line width distribution
          G(Γ) to the vesicle's number distribution N(R).
          See reference (6).

such a dramatic temperature dependence. One can imagine a
simple model to help understanding why the vesicles we studied
fuse and when they stop fusing as will now briefly be exposed.

II. B. PORE MODEL

     To explain the variation in vesicle's permeability with
osmotic pressure (7) De Gennes has proposed the following pore
model (see fig. 6).

     The energy E(r) needed to create in the vesicle's wall a
pore of radius r includes:

     - a positive contribution $2\pi r\xi$ resulting from the contact
energy between the chains and the water molecules all around
the hole. Note that this term has the same origin as the γa term
in equ.(1) except for the fact that in the present case there are
no polar heads at the interface water-hydrocarbon.

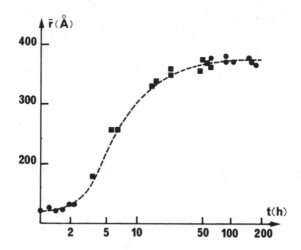

Fig. 5.   Average radius of a vesicle's sample as a function of
          time. The sample was first maintained at 24°C (dark
          circles ●) for several hours, then heated to the constant
          temperature 35.8°C (dark squares ■) for several days then
          cooled down again at 24°C for a few more days (●).

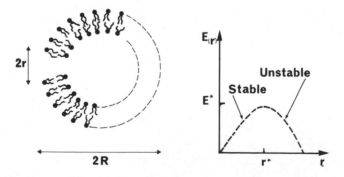

Fig. 6.   Illustration of the pore model.
          As the energy E(r) needed to create a pore of radius r
          goes through a maximum at r*, pores of radius r<r* will
          spontaneously close down (stable pores) whereas pores
          of radius r>r* will open up (unstable pores).

- a negative contribution $-\sigma\pi r^2$ arising from the surface
tension σ of the vesicle's wall, which may have several origins.
It can be due to a difference in osmotic pressure between the
inside and outside of the vesicle which we have tried to avoid
in our experiments. It can also come from built in stresses
which we believe were introduced upon cooling down the vesicles
from their sonication temperature above T_ϕ to the lower temperature
(below T_ϕ) at which they were studied.

The energy $E(r) = \xi 2\pi r - \sigma\pi r^2$ exhibits a maximum $E^* = \pi\xi^2/\sigma$
for $r^* = \xi/\sigma$. Any given pore of radius $r>r^*$ will thus tend to
open further so as to reduce $E(r)$. If one assumes the creation
of such an unstable pore to be a necessary step in the fusion
process, one can estimate the probability of such an event by
comparing E^* to kT. The crucial parameter in E^* is the surface
tension σ which we have evaluated with the following assumptions.

In a vesicle formed above T_ϕ, the polar heads occupy their
optimal survace a_o and the bilayer thickness is d (see fig.1).
As the vesicle is cooled down well below T_ϕ, the bilayer thickness
will increase to d' but the polar heads won't be able to occupy
their optimal free surface a_o' as the numbers of amphiphiles on
the outer and inner surfaces respectively are supposed to remain
unchanged (the rate of lipid exchange between the outer and inner
layer, the so-called "flip-flop rate" is typically a few hours in
the liquid state and becomes much longer in the cristalline
state). The resulting surface tension can thus be evaluated
(using equ.(3) as a function of the grometrical parameters of the
bilayer above and below T_ϕ, and of R_m, average between the internal
and external radius of the vesicle above T_ϕ:

$$\sigma = 16\pi\gamma\,d^2/R_m^2\left[(d'/d)\ \sqrt{a_o/a_o'}\ - 1\right]$$

which numerically yields $\sigma \simeq 140$ erg/cm^2.

Taking for ξ the value 10^{-6} dynes used to fit the permeability
data on vesicles (7) we obtain $E^* \simeq 0.014$ e.v which is about half
the value of kT. The opening of an unstable pore is therefore a
very probable event in the stressed vesicles we studied. However,
as soon as two or more vesicles have fused, σ will decrease and
eventually change sign (still neglecting the flip-flop effect)
and the concept of an unstable pore thus disappears. The fusion
process will then stop once all the smallest vesicles will have
fused, forming larger structures composed of 2, 3, etc. initial
vesicles. This explains why the vesicles size distribution
broadens and shifts towards larger sizes.

This very qualitative picture is only relevant far from the
phase transition temperature T_ϕ and can in no way account for the
increase in fusion kinetics as one approaches T_ϕ. However, it points

out the fact that unstable pores may spontaneously form at room temperature in the stressed vesicles we have studied, therefore explaining their unstability.

The fusion phenomenon will stop once the surface tension σ will be released, i.e. once $E^* \gg kT$.

To summarize, we have reviewed simply the basic theoretical steps explaining the formation and size distribution of vesicles. These predictions were checked by our light scattering meaurements which furthermore showed an evolution in time of that size distribution. This phenomenon was attributed to the unstability of the initial vesicles. In an attempt to explain the fusion process, we then presented a simple pore model which showed how this unstability was linked to the existence of a surface tension. In our case, this tension was built in during the preparation procedure, but it could also be introduced by other physical or chemical agents, which are known to induce fusion.

REFERENCES

1. C. Tanford, "The hydrophobic effect", Wiley (1973).
2. J. N. Israelachvili, D. J. Mitchell and B. W. Ninham, J.Chem. Soc. Far. Trans. II, 72:1525 (1976) and BBA 470: 185 (1977).
3. D. Chapman, R. M. Williams and B. D. Ladbrooke, Chem. Phys.Lipids, 1:445(1967);B.D. Ladbrooke and D.Chapman,Chem.Phys.Lipids 3:304(1969).
4. J. A. McCammon and J. M. Deutch, Jour. Am. Chem. Soc., 97:23 6675 (1975).
5. C. H. Huang, Biochem., 8:344 (1969).
6. P. Bezot, N. Ostrowsky and C. Hesse-Bezot, Optics Com., 25:14 (1978).
7. C. Taupin, M. Dvolaitzky and C. Sauterey, Biochem., 14:4771 (1975).

LIGHT SCATTERING BY WATER IN OIL MICROEMULSIONS

A. M. Cazabat and D. Langevin

Laboratoire de Spectroscopie Hertzienne de l'ENS
24 Rue Lhomond 75231 Paris Cedex 05
France

ABSTRACT

We have studied both the intensity and the autocorrelation function of the light scattered by water in oil microemulsions. A dilution procedure was ruled out, which allowed us to extract from the experimental data, information about droplet sizes and interaction forces.   The data for the variation of diffusion coefficient with droplet concentration were interpreted using recent theoretical analysis of macromolecular diffusion (Felderhof and others).

We have studied different systems, mixtures of water with or without salt, sodium dodecylsulfate (surfactant), toluene or cyclohexane (oil), butanol or pentanol (cosurfactant). It was found that in most cases the droplets do not behave like hard spheres, and that supplementary interactions must be taken into account.

I.   INTRODUCTION

Since their discovery in 1943 by Shulman and coworkers (1) microemulsions have attracted much interest. Their structure has been studied extensively by a great number of techniques X-ray diffraction (2), neutron scattering (3), light scattering (4-8), sedimentation (9, 3), electron microscopy (10, 11). It is now currently admitted that these systems are dispersions of small water (or oil) droplets surrounded by surfactant molecules forming a micelle-like aggregate. The droplet sizes are of the order of 100Å. The dispersing medium is a continuous phase containing the oil (or the water in the case of oil in water microemulsions).

As usual micelles, the droplets are not permanent aggregates. When two droplets collide, they exchange either single constituents molecules, either fractions of their membranes and water contents. The characteristic times of these processes are below 10^{-7}s, as measured by ultrasonic relaxation (12), RPE (13) and RMN (14). The diffusion coefficient D is of the order or 10^{-7}cm^2/sec. If we adopt a value of 500 Å for the interparticular distance, the diffusion time between two collisions will be of the order of 10^{-5}s, much larger than the exchange times. Therefore the brownian motion of these particles can be analysed in terms of classical diffusion processes (15). In particular the light scattering data can be related to the osmotic compressibility and to the friction force between droplets and continuous phase. In the case of ordinary spherical particules dispersions, the extrapolation of these quantities to zero particle concentration allows the determination of molecular weight and of hydrodynamic radius. At larger but moderate concentration the virial coefficients can be deduced and give information about interparticular forces.

The droplets concentration cannot be varied simply by adding oil to the microemulsion. Indeed the continuous phase may contain alcohol, and, to a lower extent, water. Adding simply oil changes the droplet size also. In order to interpret the light scattering data, the right composition of the continuous phase must be determined. If this dilution procedure is not respected, the virial coefficients and extrapolations to zero concentration can be completely meaningless.

Information about interaction forces in microemulsions are very few at the present time. A better knowledge of these forces could help to understand why microemulsions are stable systems contrary to emulsions which after a sufficiently long time separate into two phases. An approach to the interaction problem is due to A.Vrij and coworkers (5). They introduced an interaction potential sum of a hard sphere potential and a small supplementary attractive term treated as a perturbation. This model was used to fit light scattering intensity data. However it was not possible to extract from these data informations about the exact nature of the attractive potential.

In principle the knowledge of the interaction potential allows to predict the concentration dependence of diffusion coefficient, which can also be extracted from light scattering data. Several theories have been worked out recently for this purpose (16-21). Let us point out that microemulsions are simple systems to test these theories. The droplets sizes R are much smaller than the wavelengths $2\pi/k$ of the concentration fluctuations : $kR \ll 1$. Moreover, strong, long range electrostatic repulsions for which theoretical calculations become extremely complex are not effective in these hydrocarbon media.

Light scattering appears therefore as a very useful tool to study both sizes and interactions in microemulsions. Our purpose will be limited here to the understanding of the properties of moderately concentrated systems (volume fractions less than 10%). The microemulsions can be concentrated to much larger volume fractions (50%). These systems have still low viscosities ($\sim$1cp) and stay transparent. But in this concentration range, the nature of the medium is still almost completely unknown.

In the following, we will first describe sample preparation § II, then the experimental procedure § III and data analysis § IV In § V we will discuss the results.

II. SAMPLES PREPARATION

The samples are quaternary mixtures of water (bidistilled), sodium dodecyl sulfate (SDS, purchased from BDH), cyclohexane, toluene or carbon tetrachloride as oil and pentanol or butanol as cosurfactant. The last compounds are Merck products, pure for spectroscopy.

We prepared different series of microemulsions varying either the water to soap ratio in droplets, either the amount of salt in water, either the oil and cosurfactant. The composition of the continuous phase and of the droplets has been determined for each series using the dilution procedure described in ref (22) (see also ref (7)). The composition of these microemulsions is indicated in table 1. Temperature is $20^{\circ}C \pm 1^{\circ}$. The volume fraction of the droplets can be then defined as:

$$\phi = \frac{V_S + V_W^M + V_A^M}{V_0 + V_A + V_S + V_W} \qquad (1)$$

where V_0 is the oil volume, V_A^M the alcohol volume in the droplets, V_A the total alcohol volume; V_S is the soap volume calculated by dividing the soap mass by its density $\rho = 1.16 g/cm^3$ (the mixture is ideal); V_W^M is the water volume in micelles and V_W is the total water volume.

The consistency of the dilution procedure can be checked during light scattering experiments. Indeed, we will see later on in § III that the ratio ϕ/I, where I is the scattered intensity, should be a linear function of ϕ, at least for small volume fraction : $\phi \lesssim 5.10^{-2}$, if the droplet size is constant. A more complete proof of the constancy of micellar size can be found in the neutron scattering experiments, which give also the gyration radius of the droplet for each concentration ϕ. Such experiments have been done by C. Taupin and coworkers on microemulsions A, B and C (3) (see table 1).

TABLE I

Microémulsion	NaCl molarity (in water)	Droplets composition			composition of the continuous phase (in cm³)					
		SOS weight	V_M^M (cm³)	V_A^M (cm³)	C_6H_{12}	$C_6H_5CH_3$	CCl_4	$C_5H_{11}OH$	C_4H_9OH	H_2O
AX	-	0.4g	0.4	0.2		100			12,3	0.34
A	-	"	0.5	0.4	100			9.5		0.18
AR	0.1 M	"	"	0.5	.			8		0.15
AS	0.5 M	"	"	0.4	.			5.1		0
AT	1 M	"	"	.	.			4		0
AY	-	"	"	0.17		100			13.5	0.4
AZ	0.5 M	"	"	0.36		.			7.8	-
AM	1 M	"	"	0.28		.			6.4	0
A 3	-	"	"	0.19			100		12,3	0.4
B	-	"	1	0.20	100			19		0.32
BR	0.1 M	"	"	0.36	.			11		0.15
BS	0.5 M	"	"	0.30	.			4.6		0
C	-	"	1.5	0.1	.			40		0.5

III. EXPERIMENTAL PROCEDURE

As it was mentioned in § I, the droplets brownian motion is not greatly affected by droplets finite lifetimes. Therefore light scattering arises from concentration fluctuations of the droplets in the continuous phase. If <u>droplet size</u> is assumed to be <u>constant</u>, the excess scattering of the particles over that of continuous phase is (23):

$$J = 2\pi^2 \; n^2 \; (\frac{\partial n}{\partial \phi})^2 \; \frac{1}{\lambda^4} \; KT \; \phi \; (\frac{\partial \Pi}{\partial \phi})^{-1}$$

λ is the light wavelength in vacuo, n the refractive index of the microemulsion, Π the osmotic pressure, k the Boltzman constant, T the absolute temperature.

We perform relative intensity measurements at a fixed scattering angle $\theta = 90^\circ$. We do not obtain J directly, but a quantity I proportional to $J\Delta\Omega$, where $\Delta\Omega$ is the solid angle of detection. $\Delta\Omega$ varies with the refractive index of the sample as $\Delta\Omega \propto n^{-2}$. Therefore:

$$I = K \; (\frac{\partial n}{\partial \phi})^2 \; \phi \; kT \; (\frac{\partial \Pi}{\partial \phi})^{-1} \qquad\qquad (2)$$

where K is a constant, independent of the sample.

Calibration was made using polystyrene latex spheres, from Dow Chemical, as described in ref (7).

The autocorrelation function of the scattered light was also recorded on the same samples. It is given by

$$g^{(2)} \; (\tau) = 1 + e^{-2DK^2\tau}$$

where $K = \frac{4\pi n}{\lambda} \sin \frac{\theta}{2}$ and D is the diffusion coefficient. D is related to osmotic compressibility by (16)(18):

$$D = V \; \frac{\partial \Pi}{\partial \phi}/f \qquad\qquad (3)$$

where V is the volume of the droplet, f the friction coefficient between droplet and continuous phase.

Samples are put into 1cm path length glass cells (Hellma). The light source is an Argon ion laser Coherent Radiation CR2 at a wavelength 5145 Å giving an output power of about 0,5 watt. Refractive indices have been measured using an Abbe refractometer. The autocorrelation function was recorded with an analog real time correlator of 100 channels, built in the laboratory. The signal to noise ratio is excellent; Correlation times are measured within

5%. Except for some particular cases (§ IV), $g^2(\tau)$ has been found exponential in all the concentration range and we found no variation of D with scattering angle for $\theta > 4°$. The main source of dispersion of light scattering data (both I and D) arises from small unavoidable differences between sample compositions.

IV. DATA ANALYSIS

In the limit of very small volume fraction, the osmotic pressure can be approximated by (24)

$$\Pi \sim \frac{kT}{V} \phi \ (1 + \frac{B\phi}{2} \ ) \qquad V = \frac{4}{3} \ \pi R^3$$

where V is the volume occupied by the constituents of the droplets as defined in § II (see eq. 1), R the corresponding radius. $\frac{B}{2}$ is the virial coefficient. We therefore deduce from eq. 2 that:

$$\frac{\phi}{I} = K^{-1} (\frac{\partial n}{\partial \phi})^{-2} \frac{3}{4\pi R^3} \ (1 + B\phi) \tag{4}$$

As mentioned in § II, ϕ/I should be a linear function of ϕ at small ϕ if R remains constant.

Similarly, the diffusion coefficient can be written as (10)(11)

$$D \sim D_0 \ (1 + \ \alpha\phi) \qquad\qquad D_0 = \frac{kT}{6\pi\eta R_H} \tag{5}$$

η is the viscosity of the continuous phase, R_H the hydrodynamic radius of the droplet.

α is related to the virial coefficient B from eq. (3) :

$$\alpha = B - \beta$$

β being defined by friction coefficient f variation at small ϕ:

$$f \sim 6\pi \ \eta \ R_H \ (1 + \beta\phi) \tag{6}$$

Sedimentation experiments also allow to determine a friction coefficient f_{sed} (24) which is related to f through :

$$f_{sed} = f(1 - \phi) \tag{7}$$

the factor $1 - \phi$ arises from the relative motion of particles with respect to solvent in the case of sedimentation experiments. We have represented in fig. 1 the variation of ϕ/I versus ϕ for $\phi \lesssim 0.1$ and for different microemulsions. One can see that the linearity of the curves is satisfactory, indicating that the droplets

radius remains constant (in fact ϕ/I should be still a linear
function of ϕ if R increases also linearly with ϕ. But in such
case the dilution curves should not be simultaneously linear (7).

Figures 2, 3 and 4 show examples of intensity and diffusion
coefficient variations versus ϕ for different microemulsions. The
first step of data analysis has been to extrapolate ϕ/I and D to
zero ϕ. In this manner we can determine the droplet radius R and
the hydrodynamic radius R_H (eq.4 and 5) (the continuous phase
viscosity has been measured with a capillary viscosimeter). The
corresponding values have been reported in table II. In this table
we have also reported the radius of droplets aqueous cores R_W,
calculated as:

$$R_W = R \left(\frac{\phi_W}{\phi}\right)^{1/3} \qquad\qquad \phi_W = \frac{V_W^M}{V_O + V_A + V_W + V_S}$$

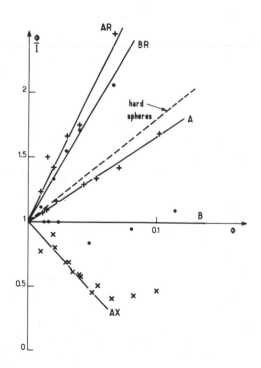

Fig. 1.   Variation of ϕ/I versus ϕ for $\phi \lesssim 0.1$, and for different
          microemulsions of table I. ϕ/I has been mormalized to 1
          for $\phi = 0$.

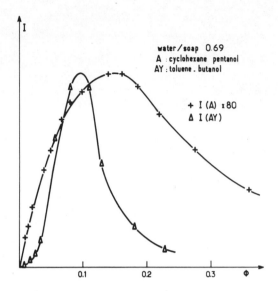

Fig. 2.   Relative intensities versus ϕ for microemulsions relative
to same water to soap ratio in the droplets, but to
different continuous phases.

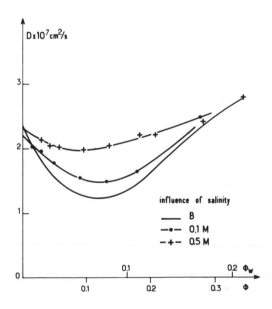

Fig. 3.   Influence of salinity on diffusion coefficient.   Full
line curves are an average of repeated experiments.

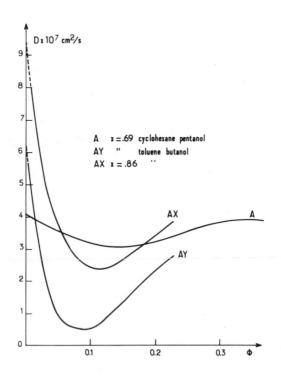

Fig. 4. Influence of water/soap ratio and of continuous phase
on diffusion coefficient. Full line curves are an
average of repeated experiments.

148 — A. M. CAZABAT AND D. LANGEVIN

TABLE II

Microemulsion	R	R_W	R_{HS}	R_H	B	A	B'	α	$α^{th}$	β
AX	29Å	21 Å	28 Å	36 Å	-12	-20	-13	-20	-8	7.5
A	49	36	48	48	6.8	-5	2.8	-3.5	-0.5	10
AR	50	35	53*	49	20	0	9.6	-3	2	~20
AS	48	34	51*	48	20	0	9.6	-1	2	~20
AT	48	34	51*	48	20	0	9.6	-1	2	~20
AY	~33	~26		41	~<-12			~<-20		
AM	34	26	33	44	0			-6		4
A3	57	45		55	-4			-20		
B	75	65	80	81	0	-15.5	-6	-11.5	-6	10
BR	84	70	89*	93	14	0	9.5	-4.5	2	~20
BS	93	77	98*	94	17	0	9.5	-2	2	~20
C	92	84	105	96	-2	-19	-6.5			

The second step has been to analyse the variation of ϕ/I and D versus ϕ at small ϕ : $\phi \lesssim 0.1$. In the limit of linear variations, the virial coefficients B, α, and β (eq.4, 5, 6) can be extracted from the data. The corresponding results are reported on Table II. The linearity of ϕ/I and D versus ϕ is limited to much smaller ϕ than for the friction force f. This means that the minimum of D versus ϕ is largely accounted for by the maximum of I versus ϕ through eq.(3).

In a last step we tried to find a form of osmotic compressibility accounting for the maximum of I versus ϕ (Fig.2). Such feature is already found in a hard spheres system for which (25)

$$\Pi_{HS} = \frac{kT}{V_{HS}} \phi_{HS}(1+\phi_{HS}+\phi_{HS}^2-\phi_{HS}^3)/(1-\phi_{HS})^3 \qquad (8)$$

ϕ_{HS} being the volume fraction occupied by spheres of radius R_{HS}, $V_{HS}=\frac{4}{3}\pi R_{HS}^3$. This system shows a maximum scattered intensity at $\phi_{HS}\approx0.13$ close to our experimental values (fig.2). However such expression does not lead to a very good agreement with experimental data. In particular as $\phi \sim \phi_{HS}$, B should be close to 8, the hard spheres value. This is not the case as shown in table II. A best fit is obtained (5) adding to the hard sphere term a small perturbation

$$\Pi_A = \frac{kT}{V} \; \frac{A}{2} \; \phi^2 \tag{9}$$

with:

$$A = \frac{4\pi}{kT} \; \frac{1}{V} \; \int_{2R_{HS}}^{\infty} V_A(r) \; r^2 dr \tag{10}$$

and $V_A(r)$ is the perturbation to hard sphere potential $V_{HS}(r)$ (total interaction potential is $V(r) = V_{HS}(r) + V_A(r)$).

In this approximation, we neglect interactions between three or more droplets, which should correspond to terms of order ϕ^3 and higher in Π_A. This model will be therefore very crude, since intermicellar distance is of the order of $2R\phi^{-1/3}$, i.e. as small as 4R at $\phi \sim 0.1$. The virial coefficient becomes within the model:

$$B' = 8(\frac{\phi_{HS}}{\phi}) + A \tag{11}$$

Examples of the fit are shown on fig.5 and 6. It is rather poor for the fig.6 and such behaviour is systematic for microemulsions containing salt. We believe that in these cases the ϕ^3, ϕ^4 ... terms omitted in eq.9 become important, indicating an increase of the interaction ranges. When this is the case the results of the fit reported on table II are followed by a star.

Even when the fit quality is good as in fig.5 one sees in table II that there are systematic differences between virial coefficients deduced from the slope of ϕ/I (B, eq.4) and from the above model (B', eq.11). Such discrepancies again indicate over-simplification of the osmotic term Π_A.

Finally we tried to interpret the initial slopes of diffusion coefficient versus ϕ. As it has been mentioned before recent theories of macromolecular diffusion allow the calculation of this slope, provided the interaction potential is known. If particles interact as hard spheres:

$$D = D_0 \; (1 + \alpha_{HS} \; \phi_{HS})$$

where $\alpha_{HS} \sim 2$. It is therefore again clear that microemulsions do not behave like hard spheres, since the values of α from table II are all negative.

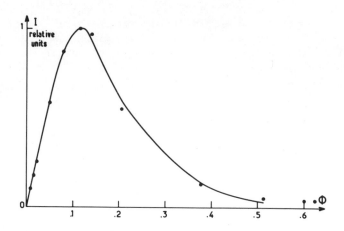

Fig. 5.   Relative intensity for microemulsion B versus ϕ.
Experimental points, - theoretical fit with eq. (2), (8)
and (9).

In a previous paper we made a comparison between our results
and the theory of Anderson and Reed (17). The agreement was very
satisfactory. However it is known now that they omitted in their
treatment several terms in hydrodynamics interactions (21). In
this paper we have compared our data to the more complete treatment
of Felderhof (21). We have used

$$D = D_0 \ (1 + \alpha\phi)$$

$$\alpha \underset{\sim}{} \alpha_{HS} + A - \frac{4\pi}{kT} \ \frac{R_{HS}}{V} \ \int_{2R_{HS}}^{\infty} V_A(r)r\,dr \qquad (12)$$

In this expression, a first contribution comes from osmotic
pressure. The second term is the contribution of hydrodynamic
interactions coming from the Oseen tensor. In Felderhofs' paper
there are other shorter range hydrodynamic contributions, which
happen to be negligible here in view of our experimental accuracy:
$\Delta\alpha \sim 1$.

It is possible to deduce α from A by making a very rough
approximation. If the range of V_A is very short,

$$\int_{2R_{HS}}^{\infty} V_A(r) \ r^2\,dr \ \underset{\sim}{} \ 2R_{HS} \int_{2R_{HS}}^{\infty} V_A(r)r\,dr$$

and consequently

$$\alpha \underset{\sim}{\sim} 2 + \frac{A}{2} \tag{13}$$

(If $V_A = - \dfrac{a}{r^6}$, we deduce from eq.10 and 12 : $\alpha = .63$ A, instead of .5A (eq.13)).

The corresponding values are shown on table II. In spite of large systematic differences: theoretical values are always larger than experimental ones by several unities, a qualitative agreement is found with the theory. A better knowledge of the interaction potential shape $V_A(r)$ should certainly improve the agreement, because in eq.13 α is overestimated.

As it has been mentioned before, D is independent of scattering angle in most cases. However, in the case of microemulsion AY which exhibits an extremely strong variation of D versus ϕ (factor $\sim$ 15 between $\phi=0$ and $\phi\sim0.1$ close to the minimum), we found also a visible variation of D with θ (factor up to 2 between $\theta=90^\circ$ and forward direction) close to $\phi\sim0.1$. We attribute this effect to droplets clusters aggregates which sizes become comparable to K^{-1}. Indeed, in such cases, the correlation time is no longer proportional to K^2 (19)(26). Note that in this system the virial coefficient is very negative, indicating strong attractive inter-action between droplets.

V.   DISCUSSION

In the following we will try to analyse the effect of the different parameters on microemulsion composition in the view of the results reported on table I and II.

Ratio x of soap to water in the droplets

This can be analysed by comparing microemulsions A, B, C with cyclohexane or AX and AY with toluene. In the two cases we found that microemulsions can be formed within a limited range of x values. It seems not possible to obtain droplets water core radii below 20Å, probably due to steric effects of polar heads of SDS molecules (in particular SDS does not form micelles in non aqueous solvents). The droplet size increases with increasing x and this increase is accompanied by a decrease of virial coefficient : the interactions become more attractive. Above a given radius $R\sim100$Å for C_6H_{12} and $R\sim40$Å for toluene, the attraction is large, stability is reduced and it becomes difficult to define a dilution procedure.

For microemulsions B and C, R < R_{HS}. This indicates that

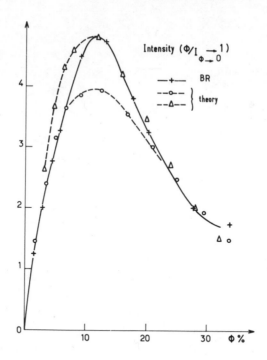

Fig. 6. Relative intensity for microemulsion BR versus ϕ
o experimental points; ———average of repeated experiments
---- theoretical fits; O best fit for the wings; the
corresponding values of R_{HS} and A are reported in Table II;
Δ best fit at the peak; this fit gives R_{HS}=90Å and A=-7.
(α_{th}= -1.5).

the volume fraction ϕ(eq.1) has been underestimated and that some
fraction of oils present in the interfacial region.

    The increase in size associated with the decrease of the soap
to water ratio in droplets ensures reduced variations of the area
per polar head of soap molecules (3). These effects are also
observed although less precisely when the sizes in toluene
microemulsions increase.

    As sizes increase, the interactions become more attractive.
However this effect is accompanied by a composition variation of
the continuous phase, which also affects the nature of interactions
(see below).

Salinity

    We have compared microemulsions A, AR, AS, AT or B, BR, BS or
AY and AW. The general features are the same:

The droplet size increases slightly, and this is associated with an increase in alcohol amount in the interfacial region. As the ratio soap to water is constant this means that the area per polar head decreases as it can be expected from the screening of electrostatic effects in the presence of salt.

The interactions become less attractive, effect associated with a decrease of the alcohol amount in the continuous phase. This seems to confirm that the nature of interactions is more sensitive to the polarity of the continuous phase than to the droplet size.

Continuous phase composition

Let us first compare microemulsions A (cyclohexane-pentanol), AY (toluene-butanol) and 3A (CCℓ_4 - butanol). The overall droplet sizes are all around 50Å.

The polarity of the continuous phase decreases from AY to A as shown by a decrease of solubilized alcohol. This is associated with a decrease of the attractive character of interactions, and with a decrease of the amount of oil in the interfacial region as already mentioned.

This suggests, as pointed out by A. Vrij (5) that an important contribution to V_A is due to the exchange of solvent molecules by segments of the soap chains when the droplets surfaces approach each other. As the polarity of the solvent increases, the hydrophobic chains are more difficult to separate and as a consequence a net attraction prevails (B < 8). If the polarity of the solvent decreases, electrostatic repulsive interactions can become important as probably for salted microemulsions (larger B). Let us point out however that these repulsions are small effects since the droplets charges are small ($\sim$ one electronic charge)(27).

CONCLUSION

Light scattering techniques appear as a very useful method to obtain informations about sizes and interactions in a microemulsion.

A very simple model for interaction forces has been used, with only two parameters: a hard sphere radius R_{HS} and the amplitude A of a small perturbation to hard sphere potential. In most cases (no salt added) the agreement with intensity data is good. Improvement of the model seems possible, but the determination of R_{HS} and A are not expected to be fundamentally different.

The diffusion coefficient data have been compared with existing theoretical treatments involving the interaction potential.

This allows us to account qualitatively for the slope of diffusion coefficient at small volume fractions. Unfortunately the theory is not applicable at larger values where D shows a minimum. However this phenomenon can be related to the maximum of intensity rather well accounted by the model that we adopted.

In the microemulsions studied here, it can be concluded that droplet sizes depend mostly on the ratio of soap to water, and interactions on the polarity of the continuous phase. A deeper understanding of the whole results will require now a better knowledge of the nature of the interaction potential.

Acknowledgements

We are greatly indebted to W. Heiss, H. M. Fijnaut, J. Rouch and C. Vaucamps for many helpful discussions.

References

1. T. P. Hoar, J. H. Shulman, Nature 102:152 (1943).
2. J. H. Shulman, D. P. Riley. J. Coll. Sci., 3:383 (1948)
3. M. Dvolaitzky, M. Guyot, M. Lagües, J. P. Le Pesant, R. Ober, C. Sauterey, C. Taupin, J. Chem. Phys., 69:3279 (1978).
4. J. H. Shulman, J. A. Friend, J. Coll. Sci., 4:497 (1949).
5. A.A. Caljé, W.G.M. Atgerof, A.Vrij in "Micellization, solubilization and microemulsions" Vol 2 Plenum Press, New York (1977).
6. A. Graciaa, J. Lachaise, P. Chabrat, L. Letamendia, J. Rouch, C. Vaucamps, M. Bourrel, C. Chambu, J. Phys. Let., 38:1-258 (1977).
7. A. M. Cazabat, D. Langevin, A. Pouchelon, J. Coll. Int. Sci. to appear.
8. S. Candau, J. Boutiller, F. Candau, Polymer, to appear.
9. J. E. L. Bowcot, J. H. Shulman, Z. Elektrochem., 59:283 (1955).
10. W. Stockenius, J. H. Shulman, L. Prince, Kolloid A., 169:170 (1960).
11. J. Biais et al CRAS, 285:213 (1977).
12. R. Zana Private Communication.
13. C. Taupin Private Communication.
14. P. Lalanne, J. Biais, B. Clin, A. M. Bellocq, B. Lemanceau, J. Chem. Phys., 75:236 (1978).
15. P. A. Egelstaff, "An Introduction to the liquid state" Acad. Press (1967).
16. G. D. Phillies J. Chem. Phys., 60:976 (1974).
17. J. L. Anderson, C. C. Reed, J. Chem. Phys. 64:3240 (1975).
18. G. K. Batchelor, J. Fluid Mec., 74:1 (1976).
19. B. J. Ackerson, J. Chem. Phys. 64:242 (1976).
20. W. Hess, R. Klein, Physica., A85:509 (1976).

21. B. U. Felderhof, J. Phys. A, 11:929 (1978).
22. A. Graciaa, J. Lachaise, A. Martinez, M. Bourrel, C. Chambu,
     CRAS, 282B:547 (1976).
23. See for instance:
     B. J. Berne, R. Pecora, "Dynamic Light Scattering" Wiley (1976).
24. See for instance:
     C. Tanford "Physical Chemistry of Macromolecules" Wiley
     (1961).
25. B. F. Carnahan, K. E. Starling, J. Chem. Phys., 51:635 (1969).
26. P. N. Pusey, J. Phys. A, 8:1433 (1975).
27. M. Laguës, R. Ober, C. Taupin, J. Phys. Lett., 39:487-L
     (1978).

DYNAMIC LIGHT SCATTERING FROM THIN, FREE, LIQUID FILMS

J.G.H. Joosten and H.M. Fijnaut

Rijksuniversiteit Utrecht
Van 't Hoff laboratorium voor Fysische en Colloïdchemie
Utrecht, The Netherlands

I.   INTRODUCTION

Light scattering techniques have proven to be a powerful
tool in studying physicochemical properties of thin, free, liquid
films. Both, time averaged (1,2) and dynamic (3,4,5,6) light
scattering experiments have been carried out on such systems.

The films we report on in this paper consist of a thin liquid
sheet bounded by two liquid-air interfaces. The stability of the
aqueous film is caused by colloidal interaction forces. The ionic
parts of the surface active molecules, located in the interfaces,
give rise to a repulsion between the interfaces. This repulsion
is counteracted by the London-van der Waals attraction forces
between all molecules in the film and by the hydrostatic pressure.

When all forces balance we are dealing with a so-called
equilibrium film. For a general review of these forces see e.g.
Ref. 7. The range of the repulsive forces can be changed by
varying the ionic strength of the solution from which the films
are drawn. This enables us to make films with an equilibrium
thickness roughly between 10 nm and 90 nm so they constitute
systems with one dimension in the colloidal domain. Therefore
soap films are considered as geometrically simple colloidal systems
and used to test and develop theories of colloid stability.

In the early sixties Vrij (8,9) recognized the possibilities
of light scattering to measure the physico-chemical properties
of thin films. If a light beam impinges on an interface most of
the light is reflected and refracted at the interface. A small
part, however, is scattered in all directions due to surface

157

corrugations caused by thermal motion of liquid and surrounding
vapor molecules. A soap film has two liquid-air interfaces
and therefore the corrugations are not only counteracted by the
interfacial tension but also by the interaction forces reaching
across the thin film. Therefore the study of light scattering,
both time averaged and dynamic, yields information on these
interaction forces.

In this paper we will report on dynamic light scattering
experiments. This implies that we are concerned with the
dynamics of the corrugations. This dynamic behaviour can be
studied either by measuring the power spectrum or the time-
autocorrelation function of the intensity of the scattered light.

II. THEORY

The instantaneous shape of an air-liquid interface can be
described by a two-dimensional spatial Fourier analysis. Since
a film has two interfaces it will be clear, by reason of
symmetry, that for each Fourier component the film motion exhibits
two modes (figure 1): one, the so-called squeezing mode, in
which the interfaces move antiparallel with respect to the normal
(on the undisturbed interface) and one, the so-called bending
mode, (10) in which they move parallel.

Next to it there are two other types of surface waves
namely the so-called longitudinal and shear waves which originate
from fluctuations inside the surface layers. As has been pointed
out by Bouchiat and Langevin (11) shear and longitudinal waves
are of minor importance in light scattering experiments compared
to the surface curvature modes. In light scattering experiments
we are always dealing with Fourier components of wavelength large

The flow patterns of the dynamic modes

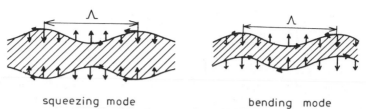

squeezing mode   bending mode

Fig. 1.   Two eigenmodes for one Fourier component with wavelength
          $\Lambda = 2\pi/q$. The amplitudes shown in the figure are
          highly exaggerated compared to the wavelength.

compared to molecular dimensions. Therefore their relaxation towards equilibrium may be described by hydrodynamic theories. For each mode, hydrodynamic theories have been developed to describe the liquid motion of the films (12-17). All these theories have in common that the liquid is assumed to be incompressible and that the nonlinear terms in the hydrodynamic equations can be neglected. The latter assumption is justified from the fact that the amplitudes of the fluctuations (≤ 1 nm) are small compared to their wavelengths (μm region). Also the effect of gravity is negligible for wavelengths of the surface waves studied in light scattering experiments (5).

There is, however, a difference in which way the effect of the interaction forces is taken into account by the several authors. We will not go into detail on this point but refer to the original papers. If one assumes a plane wave solution for the amplitude of either the squeezing or bending mode of the form

$$a(q,t) = a_o \exp \{i(qx-\omega t)\}, \; q = \frac{2\pi}{\Lambda} \; , \tag{1}$$

where q is the wavenumber and Λ the wavelength of the mode considered, x a space coordinate, ω the complex angular frequency, t the time, i the complex unity and a_o the maximum amplitude of the mode, a dispersion relation, $\omega(q) = 0$, is found from the linearized hydrodynamic theories.

The general dispersion relations, which depend on the physicochemical properties of the film system are rather complicated, but for practical purposes it is sufficient to consider some limiting cases. Before doing this we want to stress that each mode has different components according to the different solutions of the dispersion relations. From the roots of the dispersion relation one only obtains the propagation and relaxation properties of the components. However, no information can be extracted regarding the distribution of the energy along the different components of each mode. To illustrate this we consider the dynamic behaviour of the squeezing mode. Therefore we introduce a (normalized) correlation function $\phi_1(q,t)$ of the amplitude fluctuations $a(q,t)$

$$\phi_1(q,t): \;\; = \frac{<a^*(q,o) \, a(q,t)>}{<|a(q,o)|^2 >} \; , \tag{2}$$

where the brackets denote an ensemble average and the asterisk denotes the complex conjugate.

If the dispersion relation has n solutions we can write for the correlation function

$$\phi_1(q,t) = \frac{1}{<|a(q,o)|^2>} \sum_{j=1}^{n} a_j \exp(-i\omega_j t). \qquad (3)$$

The value of $<|a(q,o)|^2>$ can be found from thermodynamic
considerations i.e. by calculating the excess free energy for the
Fourier component under consideration and equating its mean value
to $k_B T/2$, the thermal energy, in accordance with the equipartion
theorem. This in fact has been done by Vrij (8,9) for both the
squeezing and bending mode. The values of a_j, however, can only
be found from a more detailed analysis of the dynamics. We
will not present this detailed calculation of the dynamics here,
but merely give a brief sketch of the results and indicate the
main features relevant for both modes. The details will be
published elsewhere (18). The spectral density $P(q,\omega)$ of a
fluctuating quantity (in our case the amplitude of either the
squeezing or bending mode) is in the classical limit of the
fluctuation dissipation theorem $\hbar\omega \ll k_B T$, related to the
generalised susceptibility $\alpha(\omega)$ by (19)

$$P(q,\omega) = \frac{k_B T}{\pi\omega} \text{ Im } \{\alpha(\omega)\} . \qquad (4)$$

There is also a simple relation between the spectral
density and the Laplace transform of the correlation function
$\tilde{\phi}_1(q,s)$ (20)

$$P_1(q,\omega) = \frac{1}{\pi} \text{ Re } \{\tilde{\phi}_1(q,s)\}_{s=i\omega} \qquad (5)$$

where s denotes the Laplace variable. One can show (18) that
both procedures yield the same result and therefore the calculation
of $P(q,\omega)$ through eq. (5) is in accordance with the fluctuation
dissipation theorem.

A similar procedure was used by Bouchiat and Meunier (21) to
calculate the spectral density of fluctuations on the interface
of a simple liquid with infinite depth.

The actual calculation is a rather tedious procedure and
therefore only some results will be given (for details see ref.
18). For the squeezing mode we find, neglecting the loading
effect of the surrounding vapor, in the so-called long wavelength
limit (LWL, $\Lambda \gg$ h) and stick boundary conditions

$$P_1(q,\omega) = \frac{\omega_{01}^2}{\pi\omega} I \{\frac{ih^2/12\nu}{D_1(q,\omega)}\} < |a(q,o)|^2>, \qquad (6)$$

where $D_1(q,\omega) = \omega + ih^2\omega_{01}^2/12\nu$ \qquad (7)

and $D_1(q,\omega) = 0$ is the dispersion relation for the squeezing mode
in this limit and

$$\omega_{01}^2 = \frac{q^2 h}{2\rho} (\gamma q^2 - 2\Pi'), \tag{8}$$

is the solution of the dispersion relation for the inviscid
case (15). The meaning of the symbols in eq. 6, (7) and (8) is:
$\nu \equiv \eta/\rho$ is the kinematic viscosity, η is the shear viscosity, ρ is
the mass density, γ is the film tension, h is the thickness of
the film and $\Pi' = d\Pi/dh$. Π is the so-called disjoining pressure
that accounts for the effect of the interaction forces and is
assumed only to depend on h. In the cases we consider in this
paper the disjoining pressure contains two contributions: one
from the electrostatic repulsion between the two charged interfaces and one from the London-van der Waals attraction between
all molecules.

Concerning the validity of the assumptions made in the
derivation of eq. (6) we found that the neglect of the influence
of the surrounding vapor for this mode is justified (18). The
LWL is always encountered in our light scattering experiments on
soap films (5). This limit in fact also implies the assumption
made with respect to the linearization of the hydrodynamic
equations. The assumption about stick boundary conditions
means that the adsorbed molecules do not move in a direction
parallel to the undisturbed interface. Such a situation is
achieved when the tangential forces, that arise from inhomogeneities
in the distribution of the surface active molecules in the interfaces, are balanced by the hydrodynamic drag at the interface.
The quantity that takes into account the surface tension
gradients is often called the elasticity coefficient and its
limit to infinity corresponds to stick boundary conditions.
From detailed calculations (18) it follows that for our monolayers
of HDTAB the elasticity is high enough to be actually in the
limit of infinity. As one can see from (6) the power spectrum is
given by

$$P_1(q,\omega) = \frac{\Gamma_1}{\pi(\Gamma_1^2 + \omega^2)} \cdot <|a(q,o)|^2>, \tag{9}$$

with $\Gamma_1 = \omega_{01}^2 h^2 / 12\nu$.

This is a Lorentzian centered around $\omega = 0$ with full width
at half maximum of $2\Gamma_1$. So the squeezing mode turns out to
be a pure diffusive mode in this limiting case.

The correlation function for the amplitude fluctuations of the squeezing mode is obtained by Fourier transforming the expression for $P_1(q,\omega)$. One obtains:

$$\phi_1(q,t) = <|a(q,o)|^2> \exp(-\Gamma_1|t|). \tag{10}$$

The same procedure is followed for the bending mode. We obtain for the spectral density $P_2(q,\omega)$ of the bending mode

$$P_2(q,\omega) = \frac{\omega^2_{02}}{\pi\omega} \, \text{Im} \, \{\frac{-1}{D_2(q,\omega)}\} <|b(q,o)|^2>, \tag{11}$$

where $D_2(q,\omega) = \omega^2 + iq^2\nu\omega - \omega^2_{02}$, and $D_2(q,\omega) = 0$ is the dispersion relation for the bending mode in this limit (15) and

$$\omega^2_{02} = \frac{2\gamma}{\rho h} q^2, \tag{12}$$

is the solution of the dispersion relation for the inviscid case. Here $<|b(q,o)|^2>$ is the mean squared amplitude for the bending mode fluctuations.

From eq. (11) one easily calculates the spectral density (if $\omega_{02} \gg \frac{k^2\nu}{2}$)

$$P_2(q,\omega) = \frac{1}{2\pi} \{\frac{\Gamma_2}{(\omega-\omega_{02})^2 + \Gamma_2^2} + \frac{\Gamma_2}{(\omega+\omega_{02})^2 + \Gamma_2^2}\}<|b(q,o)|^2> \tag{13}$$

where $\Gamma_2 = q^2\nu/2$.

Clearly $P_2(q,\omega)$ represents a Brillouin doublet with its peaks centered around $\omega = \omega_{02}$ and $\omega = -\omega_{02}$ and full width at half maximum of 2 Γ_2. Since, as is clear from eq. (13), the bending mode is a periodically damped mode we should be careful in neglecting the loading effect of the surrounding vapor phase. The reason for this is that the mass density per unit area of the film is extremely low since the film is thin (nm region).

However, the penetration depth of the disturbances into the vapor phase is of the order of magnitude of the wavelength of the disturbance, which is in the μm region, so the mass per unit area of this vapor phase that participates in the motion extends over μm and cannot be neglected although the density difference is large. In fact, this effect turns out to be very important especially in the low film thickness region. It will be clear that the inclusion of the effect of the surrounding vapor (on the liquid motion in the film) complicates severely the

analysis.   In this paper we will not go into detail on this point
but refer to Ref.18 for the complete analysis.

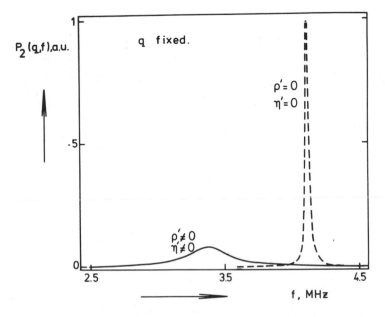

Fig. 2.   Theoretically calculated spectra for the bending mode
          with (----) and without (---) the loading effect of
          the surrounding air.   Used data: Λ = 16 µm, h = 14.5 nm,
          γ = 31 mN/m, ρ = 1000 kg/m^3, η = 1 mPa.s, ρ' = 1.2 kg/m^3
          and η' = 0.019 mPa.s.

     The importance of the loading effect is illustrated in fig.
2 where two, theoretically calculated, spectra are depicted, one
including the influence of the surrounding vapor and one according
to eq. (13).   As one can see both the frequency shift and the
width are influenced.   It turns out that the width of the peak is
mainly due to the properties of the vapor.

Summarizing we may conclude that the total spectral density of the thermally excited surface waves on a thin liquid film has features very similar to that of density fluctuations in a simple liquid. We are dealing with a kind of Rayleigh-Brillouin spectrum. The Rayleigh peak is caused by the squeezing mode whereas the two Brillouin peaks arise from the bending mode.

III. LIGHT SCATTERING

The surface waves on the thin film act as weak sinusoidal diffraction gratings. Each Fourier component acts as a grating with grating constant Λ. Since the amplitude of the component is much smaller than Λ only the first order diffraction beam is important (7,8). Besides, the specular reflection (zero order diffraction) can be used to determine the (optical) thickness of the film (22). The relation between the wave vector $\vec{q}$ of a Fourier component and the wave vectors of incident and scattered light is found from the conservation of momentum along the film plane

$$\vec{q} = \vec{q}_0^{\,1} - \vec{q}_s^{\,1} \tag{14}$$

where $\vec{q}_0^{\,1}$ and $\vec{q}_s^{\,1}$ are the projections of the incident and scattered wave vectors on the film plane, respectively. In our experimental geometry light scattered in the plane of incidence is detected and hence from (14) we have

$$\Lambda = \lambda_o / |\sin\theta_o - \sin\theta| \tag{15}$$

where λ_o is the free space light wavelength, θ_o the angle of incidence and θ is the angle of observation. The scattering from the free liquid film due to the shape fluctuations can be described by a linear superposition of squeezing and bending mode scattering. The time dependent part of the scattered field $E_s(q,t)$ is proportional to the actual amplitudes of both modes (2,8)

$$E_s(q,t) = \{Ka(q,t) + Lb(q,t)\} e^{-i\omega_o t} \tag{16}$$

where K and L are (optical) functions that depend on λ_o, θ, θ_o, h, the optical properties of the film liquid and the state of polarization and field strength of the incident laser beam and ω_o is the laser frequency. To study the dynamics of the shape fluctuations we use the technique of dynamic light scattering or intensity fluctuation spectroscopy (IFS). In the heterodyne configuration (where the scattered light is "mixed" with part of the incident light) the power spectrum or the correlation function of the photo current contains an exact replica of the field autocorrelation function (23).

So assuming that the fluctuations can be described as a stationary Gaussian random process with zero mean one obtains for the field autocorrelation function

$$\langle E_s^*(q,o)E_s(q,\tau)\rangle = \{K^2\phi_1(q,\tau) + L^2\phi_2(q,\tau)\}e^{-i\omega_o\tau} \qquad (17)$$

where τ denotes time, $\phi_1(q,\tau)$ and $\phi_2(q,\tau)$ are the autocorrelation functions of the squeezing and bending mode fluctuations, respectively. It is this function (shifted to $\omega = 0$), or its Fourier transform, that is measured in a heterodyne experiment.

In the homodyne case one detects the autocorrelation function of the scattered intensity whose fluctuating part is given by

$$|i(q,\tau)|^2 = |\langle E_s^*(q,o) E_s(q,\tau)\rangle|^2 \qquad (18)$$

$$= K^4|\phi_1(q,\tau)|^2 + L^4|\phi_2(q,\tau)|^2 + K^2L^2\{\phi_1^*(q,\tau)\phi_2(q_1\tau)+$$
$$\phi_1(q,\tau)\phi_2^*(q,\tau)\}$$

Substituting now expression (10) for $\Phi_1(q,\tau)$ and the Fourier transform of $P_2(q,\omega)$, eq. (13), in (17) and (18) we arrive at the intermediate scattering functions

$$g^{(1)}(q,\tau) = \langle I_1(t)\rangle e^{-\Gamma_1\tau} + \langle I_2(t)\rangle e^{-\Gamma_2\tau} \cos \omega_{02}\tau, \qquad (19)$$

for the heterodyne case and

$$g^{(2)}(q,\tau) = \langle I_1(t)\rangle^2 e^{-2\Gamma_1\tau} + \frac{\langle I_2(t)^2\rangle e^{-2\Gamma_2\tau}}{2}\{1 + \cos 2\omega_{02}\tau\}$$

$$+ 2 \langle I_1(t)\rangle\langle I_2(t)\rangle e^{-(\Gamma_1+\Gamma_2)\tau} \cos \omega_{02}\tau \qquad (20)$$

for the homodyne case, and where we have written $\langle I_1(t)\rangle = K^2 \langle|a(q,o)|^2\rangle$ and $\langle I_2(t)\rangle = L^2 \langle|b(q,o)|^2\rangle$.

In our experimental situation $\langle I_1(t)\rangle$ and $\langle I_2(t)\rangle$ are of the same order of magnitude (8,9,24).

From eq. (20) one sees that also in the homodyne case the frequency ω_{02} can be measured.

IV.   EXPERIMENTS

     The details of the experimental set-up are described else-
where (5). The main part consists of double walled stainless
steel vessel that can be thermostatted within ±0.002° for weeks.

     The film is formed in a (grounded) glass frame, with a hole
of about 2 cm in diameter. During all experiments there is direct
contact between the glass frame and the bulk solution.

     The film is illuminated by an Ar-ion laser about 10 mm above
the bulk liquid surface. The laser operates in the TEM_{oo} mode
(multimode) at 514.5 nm or 488 nm and at a power of $\approx$ 300 mw.
The angle of incidence is 60° and the light is polarized
perpendicular to the plane of incidence. The specularly reflected
light is measured by means of a pindiode and used to calculate
the (optical) thickness.

     The scattered light is detected by means of a photomultiplier
and the resulting voltage fluctuations of which are fed either
to a correlator or a spectrum analyzer. The correlator consists
of a A to D converter and a minicomputer programmed to calculate
the time autocorrelation function of the PMT voltage fluctuations.

     The spectrum analyzers we use are a Bruel Kjaër (BK 2007)
heterodyne voltmeter and a Hewlett Packard spectrum analyzer
(HP 3585A).

     The scattering cell and the optical system is placed on a
vibration free table in order to eliminate exterior vibrations.

     The soap solutions, from which the films are drawn, were
prepared using the cationic soap hexadecyltrimethylammoniumbromide
(HDTAB). The solutions we use contain as standard components
9.10^{-4} mole/dm^3 HDTAB plus 1 mol/dm^3 glycerol (5). To increase
the ionic strength of the solution we add KBr. This results in
a decrease of the electrical double layer repulsion and hence in
a lower equilibrium thickness. The surface tensions γ_o of the
different solutions is measured by means of a stalagmometer
(drop weight method).

V.   RESULTS

     Before a film is drawn the surface of the bulk liquid in the
glass vessel is refreshed (5). Once a film is drawn it takes
about 12 hours to reach its equilibrium thickness. During this
drainage process the intensity of the specularly reflected light
is monitored and the last maximum in this intensity is used to
calculate the thickness measurement.

This drainage process proceeds slow enough to be able to perform dynamic light scattering experiments during the thinning process and not only on equilibrium films.

A calculation of the relaxations times and frequency shifts shows that, if we use realistic values for the different parameters that enter the theory, the squeezing mode exhibits its relevant features in the ms region whereas for the bending mode this is in the μs region. For measuring the properties of the squeezing mode the homodyne detecting scheme is used with angles of observation θ between 40° and 80° with the exclusion of a small region around the reflection angle of $\theta_o = 60^o$. The exclusion of this region of some degrees is due to the gradual, but not exactly reproducible, transition from the homodyne to the heterodyne detection mode. The range of wavelengths that is embedded in this region is roughly between 2 μm and 17 μm. One can perform measurements of the relaxation time either at a fixed angle with varying thickness (during the drainage) or at fixed thickness with varying angle (equilibrium experiments). We have doen both kinds of experiments and we refer to ref. 5 for a detailed description. Therefore we will merely quote some results here. From both types of experiments we obtain a value for γ/η but from the equilibrium experiment we also obtain a value for Π/η (see eq. (9)). The experimental value of γ/η turns out to be lower than the value of this ratio we get from γ_o and η_o, the surface tension and viscosity found for the bulk solution respectively. The discrepancy is about 40% compared to the bulk values. Since we obtain the ratio of γ over η we cannot decide a priori whether the surface tension is low or the viscosity is high compared to the bulk value. It is important, however, to know separately the values of γ and η because otherwise no reliable data concerning the interaction term Π' can be deduced from the experiments.

As we have seen in the theoretical part the detection of the bending mode would provide us with a value for the surface tension of the film. Although it is not clear a priori that a film tension measured in the MHz frequency range (bending mode) is equal to the value of this quantity obtained in the kHz frequency region (squeezing mode) a measure on the bending mode could give us at least an estimate for its value.

Very recently we succeeded in detecting the bending mode by means of IFS (6). Altough the scattered intensity of the bending mode is comparable with the intensity scattered by the squeezing mode the signal to noise ratio becomes poor for large scattering angles because of the high frequencies involved. The range of scattering angles is confined to roughly $\pm5^o$ from the reflected beam depending on the film thickness. This means that we gradually move over from heterodyne to homodyne detection. Very close to the

reflected beam the scattered light is mixed with part of the
specularly reflected light that acts as the local oscillator.
Since the laser intensity noise shows its maxima well below the
actual frequencies of the bending mode we have no problems with
this noise.

As we have seen from eq. (20) also in the homodyne case
the frequency shift of the bending mode can be measured since the
squeezing mode plays the role of the local oscillator in this
case. Fig. 3 shows an example of such a homodyne spectrum.

Until now we only performed bending mode experiments on
equilibrium films but very recently we are also able to perform
measurements on draining films. Here, however, we present no results
of the experiments on draining films. So at each scattering angle the
power spectrum of the scattered light is measured. This results
in a plot of the peak frequency $f_o = \omega_{02}/2\pi$ versus wavenumber q.

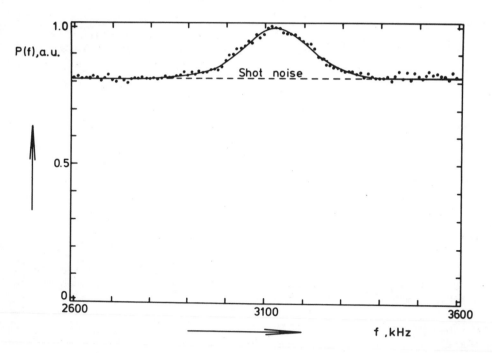

Fig. 3.   Measured power spectrum of the photo current fluctuations
          of the PMT (homodyne).

          $\Theta_o = 60^o$, $\theta = 56.5^o$, $\lambda_o = 488$ nm and h = 48.5 nm.

In fig. 4 the experimental results obtained from 4 films at
constant thickness are depicted. The solid lines in this figure
were obtained by calculating the theoretical dependence of
f_o vs. q including the loading effect of the surrounding air at
the temperature the experiments were carried out (27° C).

    As can be seen the experimental data fit very well to the
theoretical curves. The film tensions data used to fit the
experimental results are gathered in table I together with the
ionic strengths and the surface tensions γ_o of the solutions from
which the films were drawn. As one can see from this table the
tensions measured on the films differ from the bulk especially
for the thicker films. It should be realized that the film
tension values were obtained from experiments on a time scale
totally different from the time scale encountered in obtaining
the bulk values. This implies the possibility of processes that
influence both values in a totally different way like adsorption
and desorption phenomena.

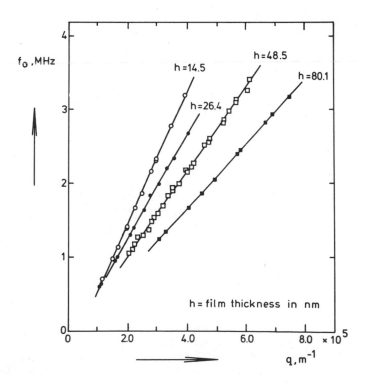

Fig. 4.   Peak frequencies $f_o(q)$ vs. wavenumber q for 4 film
          thicknesses; solid lines are theoretical curves including
          the loading effect of the surrounding air ($\rho' = 1.2$ kg/m³
          and $\eta' = 0.018$ mPa.s).

TABLE I

Bulk solution properties, equilibrium thicknesses and
film tensions of the 4 systems investigated; c means the
ionic strength of the solution (= concentration HDTAB
plus the concentration KBr).

c (mmole/dm^3)	γ_o (mN/m)	h (nm)	γ (mN/m)
0.8	40	80.1	31
1.9	35	48.5	32
10	33	26.4	31
50	31	14.5	29

At this moment, however, the discrepancy is not fully understood. Although it is known that the interaction forces can
lower the actual surface tension of the film compared with the
surface tension of the bulk solution (25,26), this lowering effect
is generally much smaller than is found in our experiments.
Furthermore this effect of the interaction forces on the surface
tension is of static nature but there may be also an influence of
the interaction forces on the dynamics of the bending mode
fluctuations. There exist more refined electrohydrodynamic
theories on the dynamics of surface waves on thin films (14,15)
than the one we used in this paper. Whether this different
theoretical approach yields results that explains our
experimental data is now under investigation.

Anyway, if one accepts the values for the surface tension of
the film obtained from the bending mode experiments one finds at
least a partial explanation for the differences found in the
squeezing mode experiments with respect to the ratio γ/η . It
turns out, however, that even if we use these relatively low
film tensions the value for the viscosity coefficient η obtained
from the dynamic light scattering experiments differs from the
bulk value η_o.

Until now we did not try to interpret the widths of the
peaks obtained in the bending mode experiments. The main reason
for this is, as we already mentioned in section II, that the width

of the peak mainly results from the properties of the vapor phase. Hence, it will be difficult to extract reliable information from the measured spectra about the damping mechanism that arises from the film itself. Another reason from a more practical point of view, is that the divergence of the laser beam and the geometry of our optical system can obscure the spectra in such a way that a difficult deconvolution procedure is necessary to obtain the real width of the peak.

We want to stress that this problem is much more important for spectra which are frequency shifted than for spectra that are centered around frequency zero like in the case of the squeezing mode.

VI. CONCLUSIONS

Summarizing we may conclude that by laser intensity fluctuation spectroscopy physico-chemical properties of thin, free, liquid films can be measured even in the MHz range. The relaxation phenomena can be described by means of linearized hydrodynamic equations. The observed behaviour of the time dependent quantities versus the wavenumber q is qualitatively in accordance with the theory. The quantitative discrepancies show up as deviations between bulk and film values of surface tension and viscosity. These differences seem to depend on the film thickness and a full explanation is not available at this moment.

ACKNOWLEDGEMENT

This work was part of the research programme of the "Stichting voor Fundamenteel Onderzoek de Materie" (FOM) with financial support from the "Nederlandse Organisatie voor Zuiver Wetenschappelijk Onderzoek" (ZWO).

REFERENCES

1. J. A. Mann Jr., J. Colloid Interface Sci., 25:437 (1967).
2. W. A. B. Donners, J. B. Rijnbout and A. Vrij, J. Colloid Interface Sci., 61:535 (1977).
3. H. M. Fijnaut and A. Vrij, Nature (London) Phys. Sci., 246:118 (1977).
4. Very recently Dr. D. Langevin (Paris) pointed out to us that also Dr. C. Young did his Ph.D. thesis on dynamic light scattering on soap films (Harvard 1977).
5. H. M. Fijnaut and J. G. H. Joosten, J. Chem. Phys., 69:1022 (1978).
6. J. G. H. Joosten and H. M. Fijnaut, Chem. Phys. Letters 60:483 (1979).

7. R. Buscall and R. H. Ottewil, Specialists Periodical Reports: Colloid Science, edited by D. E. Everett (The Chemical Society, London, 1975), Vo. 2.

8. A. Vrij, J. Colloid Sci., 19:1 (1964).

9. A. Vrij, Adv. Colloid Interface Sci., 2:39 (1968).

10. There is some ambiguity in the nomenclature in literature on the term bending mode; sometimes it is also called stretching mode (e.g. in ref. 6). We feel that it would be more appropriate to attach the term stretching mode to longitudinal fluctuations and not to curvature waves.

11. M. A. Bouchiat and D. Langevin, J. Colloid Interface Sci., 63:193 (1978).

12. J. Lucassen, M. van den Tempel, A. Vrij and F.Th. Hesselink, Kon. Ned. Akad. Wet. B 73:109 (1970).

13. A. Vrij, F.Th. Hesselink, J. Lucassen and M. van den Tempel, ibid B 73:124 (1970).

14. B. U. Felderhof, J. Chem. Phys., 49:44 (1968).

15. J. G. H. Joosten, A. Vrij and H. M. Fijnaut, Proc. Intern. Conf. Phys. Chem. and Hydrodynamics (Levich Conference), 1977 (Advanced Publications, Guernsey, 1978) Vol. 2, p.639.

16. S. Sche and H. M. Fijnaut, Surface Sci., 76:186 (1978).

17. S. Sche, J. Electrostatics 5:71 (1978).

18. J. G. H. Joosten, to be published.

19. L. D. Landau and E. M. Lifschitz, Statistical Physics (Pergamon Press, London-Paris, 1959), p.391 f.f.

20. B. J. Berne and G. D. Harp, Adv. Chem. Phys., 17:63 (1970).

21. M. A. Bouchiat and J. Meunier, J. Phys., (Paris) 32:561 (1971).

22. J. B. Rijnbout, J. Phys. Chem., 74:2001 (1970).

23. H. Z. Cummins and H. L. Swinney, Progress in Optics VIII, edited by E. Wolf (North-Holland Publ. Co., Amsterdam, 1970), p.133.

24. J. A. Mann Jr., K. Caufield and G. Gulden, J. Optical Soc. Am., 61:76 (1971).

25. H. M. Princen and S. Frankel, J. Colloid Interface Sci., 35:386 (1971).

26. J. A. de Feyter, J. B. Rijnbout and A. Vrij, J. Colloid Interface Sci., 64:258 (1978).

TECHNIQUES FOR LIGHT SCATTERING FROM HEMOGLOBIN*

Ann H. Sanders and David S. Cannell[+]

Department of Physics
University of California, Santa Barbara
Santa Barbara, CA 93106

ABSTRACT

This paper describes methods we have found sufficient to obtain accurate diffusion coefficient measurements of hemoblobin using quasi-elastic light scattering. The same methods should be applicable to other proteins or liquid samples which absorb at the laser wavelength. The results are precise enough to detect a change of approximately 0.5% in the diffusion coefficient of hemoglobin upon oxygenation.

INTRODUCTION

The purpose of this paper is to describe in detail various methods and techniques which we have found to be helpful in obtaining the most accurate possible results when making quasi-elastic light scattering measurements on hemoglobib, a protein which absorbs strongly throughout the visible region of the spectrum (1). Such absoption results in heating of the solution in and near the laser beam, and thus in convection as well. Since the viscosity of the solvent varies approximately 2% per degree centigrade, even incident powers as small as 10mW can cause significant changes in the viscosity of the solution in the immediate vicinity of the beam. Of course, such a change in viscosity effects the measured linewidth or correlation time and thus the apparent diffusion coefficient.

*This research supported by the National Science Foundation under grant NSF PCM76-21328.
[+]John Simon Guggenheim Fellow.

In principle one can obtain the correct diffusion coefficient under these conditions simply by making measurements as a function of laser power and extrapolating the results to zero power, and indeed we find this procedure to be absolutely necessary. Furthermore it is important to carry out such an extrapolation for every sample, because there is inevitably a slight amount of methemoglobin in every normal sample, and since methemoglobin is very strongly absorbing even slight variations in the amount present significantly alter the total absorption of the sample.

We have also found it to be advantageous to perform the measurements with the beam quite close to and parallel to one face of the scattering cuvette. This not only reduces the total path of the incident and scattered light in the cell, which is important in terms of achieving adequate scattered intensities, but also serves to greatly reduce the convective motion of the sample. This is necessary because in general, when convection is adequate to provide the dominant mechanism for removing the absorbed heat from the vicinity of the beam, one observes a greatly reduced dependence of measured correlation time on incident beam power. This is deceptive however, because, as the power is reduced further, conduction becomes the dominant mechanism and a much steeper slope is observed. Thus if one were to make extrapolations to zero power based on measurements made in the convective region errors of 5% or more could easily be made. On the other hand, when the scattering region is close to the corner of a square 1 cm x 1 cm cuvette we find no evidence of convective effects for incident powers as high as 25 mW, and one can routinely obtain diffusion coefficients with an accuracy of a few tenths of one percent.

The remainder of this paper consists of four sections. The first outlines the procedures we have followed to prepare pure hemoglobin from whole blood and to deoxygenate samples when desired. The second covers the filtration technique used to prepare extremely clean samples, and the various checks we have found helpful in verifying a sample's cleanliness. The third discusses the use of cylindrical focusing optics for the incident beam in order to further reduce heating effects, while the fourth and final section presents some of the results we have obtained for oxy and deoxyhemoglobin.

Preparation of Hemoglobin

We found it necessary to prepare our own hemoglobin from whole blood in order to obtain accuract and reproducible values for the diffusion coefficient. Although hemoglobin is available commercially, it is sold only in the met form in which the iron has been oxidized to the ferric (+3) state. Methemoglobin is extremely stable, but it is difficult to convert it to normal ferrous (+2)

hemoblobin. Furthermore even when stored at 4°C normal
oxyhemoglobin slowly oxidizes to methemoglobin. For this reason
one should use samples which are less than two weeks old, and
the protein should be stored at 4°C. We saw no measureable
change in diffusion coefficient over a ten day storage period after
preparation, although the concentration of methemoglobin in the
sample increased for less than 3% immediately after preparation
to approximately 5% after 10 days.

Oxyhemoglobin was prepared using the method of Rossi-
Fanelli, et al (2) as modified by Antonini and Brunori (1),
which produces very pure (95%) hemoglobin. Immediately before
the preparation, blood was drawn over Ethylenedinistrilo
Tetraacetic Acid (EDTA) to prevent coagulation. The hemoglobin
was further purified by passing it through a diethylaminoethyl
(DEAE) column equilibrated with buffer, at pH = 7.0. The resulting
solution was at approximately 28 mg/ml concentration.

We used sodium phosphate monobasic (NaH_2PO_4) buffer at o.1 M
concentration, with the pH adjusted to 7.0 by addition of NaOH.
Choice of buffer can be important because the normal oxyhemoglobin
tetramer dissociates into dimers to a certain extent, and the
amount of dissociation depends on the buffer used (3). In
particularly buffers involving chloride cause a great deal more
dissociation than phosphate buffers.

Deoxyhemoglobin was prepared by passing premoistened nitrogen
over a shallow (7 mm) layer of oxyhemoglobin solution, while
stirring very slowly for approximately 90 minutes. We suspect
that stirring can cause slight aggregation, and it should be
minimized so far as possible. Deoxyhemoglobin was withdrawn using
a syringe and a small piece of teflon tubing which was permanently
mounted in the flask used for deoxygenation. No difficulties
were encountered in maintaining the solution oxygen free during
filtering, and while in the cell, provided nitrogen had been
used for flushing all cell filling tubing, filters, etc.

Cell Cleaning and Filling

For very high accuracy work it is essential that the sample
be scrupulously clean, in the sense of being free of particulate
contamination. Such contamination takes the form of dust particles,
protein aggregates, and microbubbles. The "crossbeating" between
the light scattered by the protein, and the spectrally narrow
light scattered by such contaminants generates a signal having a
spectral width half that of the signal from the protein alone.
The resulting total spectrum has an accurately Lorentzian lineshape
(4), (i.e. an exponential correlation function), but a linewidth
which corresponds to an erroneously small diffusion coefficient.

In addition to care in sample cleaning, one can also
discriminate against the effects of particulate contamination by
monitoring the total photocurrent, and interrupting processing of
the signal whenever the photocurrent amplitude exceeds a preset
value. This value is determined by the average photocurrent and by
the fluctuations in the photocurrent. It should ideally be equal
to the average plus several rms deviations of the photocurrent.
In this way the current level is very unlikely to exceed the
preset level except when a particle enters the illuminated
volume.

Although we were prepared to use such a discrimination
technique, it proved to be completely unnecessary because the
cell cleaning and filtering procedure outlined below produced
samples for which the photocurrent exhibited only the expected
statistical fluctuations for periods as long as 18 hours after
filling the cell.

The cell filling apparatus is shown schematically in Fig. 1.
Before any sequence of measurements, the entire apparatus was
disassembled and cleaned in Chromerge in an ultrasonic cleaner
for several minutes. After repeated flushing with distilled
water the components were cleaned ultrasonically in distilled
water for another minute. After loading with a 0.22μm
Millipore filter, the dry filter holder and connecting tubing
were further cleaned by passing nitrogen at 1 psig through the
filter for 30 minutes. After being loaded in the holder, the 0.22μm
wet filter was flushed with 60 cc of boiling water to remove any
surfactants which might denature the protein and cause aggregation.
This precaution was taken routinely with all filters used for
water, buffer or protein solution.

The apparatus was then assembled, and the cell was filled
with Glacial Acetic Acid via the filter bypass tube, and cleaned
ultrasonically for 1 minute. The acid was blown out by inverting
the cell and admitting air through the dry filter using a syringe.
The cell was then filled with distilled water which had been
prefiltered through a 0.22μm Millipore filter, and again emptied.
The process of filling the cell with water and emptying it was
repeated 5 times.

Finally, the cell was flushed using 40 cc of prefiltered
distilled water, which usually resulted in an extremely clean
cell. At this point cleanliness was easily judged by examining
the cell filled with water in a 500 mW unfocused Argon Ion
laser running at either 5145 Å or 4880 Å. The beam in the cell
was examined at a $90°$ angle using a 20X microscope, and at $\sim 15°$
scattering angle using the unaided eye. A clean cell showed no
observable sparkling objects in the beam by either method of
observation.

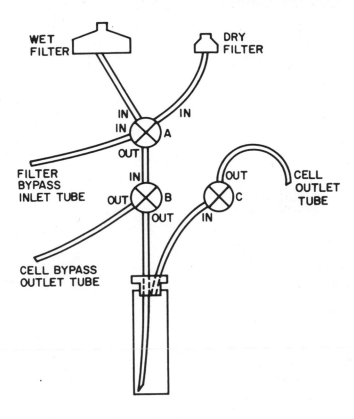

Fig. 1. Cell filling apparatus.

If judged to be clean the cell was emptied, flushed with
20 cc of prefiltered buffer, and again emptied. The protein
solution was then admitted to the cell, taking care to discard
the buffer which remained in the wet filter by means of the
filter bypass tube. When the entering protein solution reached
valve B the valve position was changed from the bypass position
to the cell filling position, and the protein entered the cell.

In the case of deoxyhemoglobin, all steps were identical
except that the cell was filled with deoxygenated buffer, and
emptied using nitrogen to force the buffer out. In addition
the first few cc of protein solution were discarded via the filter
bypass tube to avoid any possiblity of admitting accidentally
reoxygenated solution to the cell.

The final test for sample cleanliness was performed
continuously while data were being taken. It consisted of
monitoring the photocurrent using a chart recorder. As shown
in fig. 2A the photocurrent exhibited only the theoretically

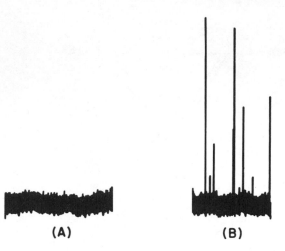

(A)   **(B)**

Fig. 2.   Photocurrent for a clean (A) and a contaminated (B)
          scattering sample. The peak-to-peak fluctuations in (A)
          are 1.6% of the dc photocurrent, which is the expected
          value. The record in (A) is for 30 minutes.

expected statistical fluctuations for periods of up to 18 hours.
In the case of a contaminated cell, the larger particles cause
sharp spikes in the photocurrent as shown in Fig. 2B.

     In addition to the requirements of a clean sample, the
mechanical arrangement used to hold and position the cell must
be adequate to ensure repeatability of the scattering angle. It
is also important to monitor or control the sample temperature
carefully since a one degree centigrade temperature change
results in a 2% error in diffusion coefficient.

Cylindrical Focusing Optics

     In order to minimize the extent to which the temperature of
any fluid element is raised by absorption of the incident laser
light, we have found it to be useful to replace the normal
spherical focusing lens with a cylindrical lens.

     In using a cylindrical lens one must be cognizant of the
effects on the spatial coherence properties of the scattered
light. In particular one must be certain that the optics used
to collect the scattered light are accepting light which has
remained in the plane defined by the sheet-like focal region of
the incident beam. Failure to observe this precaution will
decrease the amount of spatial coherence in the scattered field,
because, as viewed by the collection optics, the apparent source
dimension transverse to the focal sheet is given by

$$t_{eff} = W|\sin\alpha| + t\cos\alpha,$$

where W is the beam width in the focal region, t the thickness of the focused beam and α the angle between the axis of the collection optics and the plane defined by the beam focus, which is assumed to be a horizontal plane for convenience. Since the vertical extent of the coherence areas of the scattered light is inversely proportional to t_{eff}, the number of photoelectrons detected per coherence area per correlation time of the scattered field decreases as t_{eff} increases. Since in general this reduces the precision with which a measurement may be made in a given time, care should be taken to have $t_{eff} \lesssim 1.2t$.

The second effect which must be considered is also concerned with spatial coherence, and it limits the maximum width the beam may have in the focal region for any given focused thickness. This limitation arises because one wishes to maintain the spatial coherence properties in the scattered light despite the differently shaped focal regions of cylindrical versus spherical focusing optics. In order for this to be possible, the beam width must be small enough for the spherical waves emitted from scatterers on the opposite sides of the beam to remain accurately phase matched over the area for which one would expect spatial coherence in the scattered field when using a beam of negligible width. This requirement is illustrated by Fig. 3 which shows schematically a cross section of the focal region obtained using cylindrical optics.

The angle β is the angle over which one expects coherence in the scattered field to persist in the vertical direction and is given by $\beta \simeq \lambda/t$. Since molecules on the two different sides of the beam emit spherical wave fronts of different radii they cannot interfere perfectly even over the angle β. In order to prevent severe degradation of the angle over which coherence in the scattered field exists, it is sufficient to restrict the beam width W to the point where there is less than $\lambda/4$ total path length difference over the angle β for waves emitted from the two extreme edges of the beam. This requirement immediately leads to the restriction

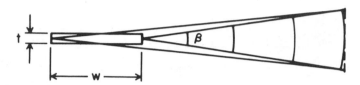

Fig. 3.   Focal region of the beam using cylindrical focusing optics.

$$W \lesssim 2\lambda(f/D)^2 \; ,$$

where f is the focal length of the cylindrical focusing lens
and D is the diameter of the beam before focusing. This is of
course simply the statement that the beam width must be no
longer than the length over which a diffraction limited focus
is attained. In practice this restriction is not a severe
limitation.

RESULTS

This section presents some preliminary results indicating
that the procedures outlined in the three previous sections
are adequate to obtain extremely accurate diffusion coefficient
measurements on hemoglobin, and that the same results are obtained
for independent preparations of the protein.

The measurements were made at a 90° angle using a 20
channel real time spectrum analyzer to measure the photocurrent
spectrum. The measured spectra were accurately Lorentzian as
shown by Fig. 4 which gives the percentage deviation between
the measured spectrum and the least mean squares fit to a
Lorentzian plus a shot noise background. The measurement time was
570 seconds, and the predetection signal to noise ratio was 3.2.

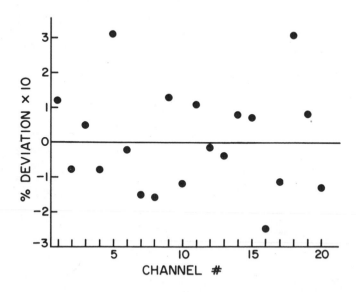

Fig. 4.   Percent deviation between the measured spectrum and the
          fit for the twenty channels.

The reduction in the dependence of measured linewidth on
incident power obtained using cylindrical focusing optics is
shown in Fig. 5. The beam width was only 1.2 mm, and thus in
principle further improvement could be made by increasing the
beam width. The measurements were made approximately 0.5 mm from
one corner of the scattering cell, and as may be seen, the
dependence of linewidth on power is highly linear permitting
accurate extrapolation to zero incident power.

Several measurements of Γ vs. power were made of oxyhemoglobin
from each of two different preparations, each measurement
involving independent cell filling, etc. The results obtained for
the extrapolated value of Γ at zero power showed a root mean square
deviation of 0.5%, and the mean values for the two preparations
were equal to within 0.03%, which is less than the expected
statistical uncertainty.

Lastly Fig. 6 shows measurements of Γ versus power for one
sample having a concentration of 13 mg/ml which was deoxygenated
and then reoxygenated. As may be seen the linewidth for the
deoxyhemoglobin is $\sim$0.5% larger than that for the oxyhemoglobin.

Further measurements are currently in progress to verify that
this change in diffusion coefficient upon oxygenation is a real
effect, and to measure the difference in diffusion coefficients
of methemoglobin and oxyhemoglobin.

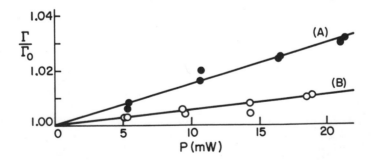

Fig. 5.   Linewidth dependence on incident laser power for
oxyhemoglobin (13 mg/ml) using (A) spherical and (B)
cylindrical focusing optics.

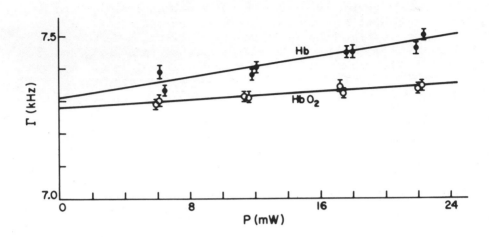

Fig. 6.   Linewidth as a function of laser power for oxy-(Φ) and
          deoxyhemoglobin (Φ).

ACKNOWLEDGEMENTS

We would like to acknowledge with deep appreciation many
helpful and stimulating conversations with Dr. Stuart Dubin who
was deeply involved in the early stages of this work, and who
suggested hemoglobin as a suitable protein for a model system
undergoing a conformational change. We also wish to acknowledge
with thanks the constant encouragement and help of Professor
Daniel Purich, whose laboratory facilities we used in preparing
and purifying the hemoglobin.

One of us (DSC) is extremely grateful to the John Simon
Guggenheim Foundation for support during the later stages of
this work.

REFERENCES

1. E. Antonini and M. Brunoir, "Hemoglobin and Myoglobin in their
    Reactions with Ligands," North-Holland Publishing Company,
    Amsterdam (1971).
2. A. Rossi Fanelli, E. Antonini and A. Caputo, J. Biol. Chem.,
    236:391 (1961).
3. E. Chiancone, N. M. Anderson, E. Antonini, J. Bonaventura,
    C. Bonaventura, M. Brunori and C. Spagnuolo, J. Biol. Chem.,
    249:5689 (1974).
4. S. B. Dubin, "Quasielastic Light Scattering from Macromolecules,"
    Ph.D. thesis, M.I.T., Cambridge, Massachesetts (1970).

QUASI-ELASTIC LIGHT SCATTERING IN THE MEASUREMENT OF THE MOTION OF FLAGELLATED ALGAE

Cesare Ascoli and Carlo Frediani

Laboratorio per lo studio delle proprietà fisiche di
biomolecole e cellule
C.N.R.-Pisa, Italy

INTRODUCTION

The aim of this communication is to show the application of quasi-elastic light scattering to the study of the motion of flagellated algae and of related-biological phenomena.

Two of these microorganisms (Euglena gracilis and Haematococcus pluvialis) are shown in fig.1. Euglena gracilis (fig.1a) is an elongated unicellular microorganism whose dimensions are 10 x 10 x 60 μm; it shows many structures, particularly photosynthetic organelles (chloroplasts) and a photoreceptor whose function is to control its phototactic response. Phototaxis is a collective motion of such microorganisms toward a light source. In Euglena a photo-kinetic response has been observed too (1). Photokinesis is a light-dependent change in the swimming speed of the cells. Haematococcus pluvialis (fig.1b) is a unicellular quasi-spherical microorganisms. It too shows a phototactic response, but the pigment that controls this photoresponse is quite unknown.

Our aim has been to obtain quantitative measurements of the mean motion parameters of a population of these algae in various experimental conditions. Until recently these measurements were carried out by microscopic observation or by the analysis of cinematographic tracks. Both techniques are much slower than laser velocimetry in measuring the statistically averaged properties of cell populations; these techniques are also less accurate, because they require the subjective selection of single microorganisms to be tracked.

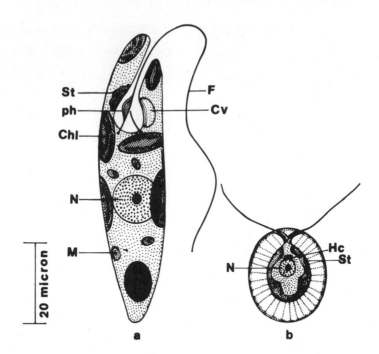

Fig. 1.   Schematic drawing of: a) <u>Euglena gracilis</u>, b) <u>Haematococcus</u>
           <u>pluvialis</u>, N nucleus, St stigma, F flagellum, Chl
           chloroplast, Hc Hematochrome, Ph photoreceptor, M
           mitochondria, Cv contractile vacuole.

      The use of laser light scattering makes it possible to study
the effect on motility of parameters such as light, temperature,
the composition of medium and so on.

      We have mainly worked on E.g. and our experiments of H.p.
cultures are still in progress. The first remark to be made is
that these microorganisms are much larger than the laser
wavelength and that they are not optically homogeneous. Thus the
light scattering pattern will not be isotropic, as shown by fig.2.
In fact, for E.g. we can observe a sharp bright peak (a whisker)
at right angles to the direction of motion and a more uniform but
weak scattered light over a large area around the axis of motion.

      The second remark to be made is that the motion of a <u>Euglena</u>,
which is illustrated in fig.3, is almost helcoidal, with a
rotation frequency, as estimated microscopically, of about 2Hz.
The flagellar beating causes the front of the cell body to
oscillate during its motion; a projection of its path on to a
plane perpendicular to the direction of motion has been drawn in
the inset in fig.3. The body rotation can be observed in the

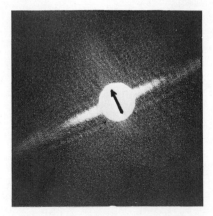

Fig. 2. Light scattering pattern from a single Euglena cell.
         Photo obtained with the film fixed parallel to the sample
         and about 20 cm from it at right angles to the laser
         beam. The arrow indicates the direction of cell motion.

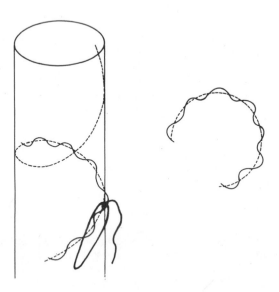

Fig. 3. Schematic drawing of Euglena motion. After Ascoli et
        al. (1978a).

light scattering pattern; in fact, the sharp peak shown in fig.2 oscillates at a frequency of about 2Hz.

These qualitative observations allow us to infer that it is very difficult to obtain quantitative results by the spectral analysis of light scattered from E.g. samples without previous knowledge of how different motion parameters act on the spectra and without selecting suitable experimental conditions.

MEASUREMENT OF MOTION PARAMETER IN ORIENTED SAMPLES

One valuable experimental approach has been that of orienting the microorganisms and choosing suitable scattering angles. Euglenae can be oriented by applying a radiofrequency field, as shown in fig.4, after resuspension in a medium of low electrical conductivity (10^{-2}mho/m). Living and dead Euglenae both orient parallel to the electric field for frequencies in the 0-10MHz range, perpendicular to it for frequencies in the 10-30MHz range, and parallel again above 100MHz (2,3). The orientation phenomenon is merely a passive electrical effect; it is the result of interaction between the dipole moment induced by the electric field in the cell and the field itself (4).

Now, by observing the scattered light pattern of oriented samples shown in fig.5, we choose suitable photodetector positions (positions α and β in fig.5) in order to measure the spectral contributions of different motion parameters separately. The measurements have been obtained by a two-beam heterodyne spectro-meter described by Ascoli et al. (1978) and also by the direct detection of scattered light. Direct detection of light scattered at right angles to the motion direction gives two spectral lines around 2 and 35 Hz, as shown in fig.6. These frequencies are independent of the scattering angle, and there are no Doppler lines because the cells are moving perpendicularly to the scattering plane, as illustrated in fig.7. These lines, which are the result of modulations of the scattered light intensity produced by the periodic movements of the cells, yield the distribution of body rotation frequencies (the 2Hz line) and flagellar beating frequencies (the 35Hz line). These intensity modulations are negligible when we collect the light scattered in the large region around the direction of motion (position β in fig.5). In fact, the line at 2Hz vanishes and the 35Hz line is greatly reduced (fig.8).

Now, in the experimental conditions shown schematically in fig.9, the swimming velocity $\vec{v}$ lies in the scattering plane $\vec{K}_0\vec{K}_S$, and is perpendicular to the laser wavevector $\vec{K}_0$. For a single swimming cell, the Doppler shift is then:

$$\Delta\upsilon = \frac{2v}{\lambda} \cos\alpha \sin \frac{\theta}{2} = \frac{v}{\lambda} \sin \theta \qquad (1)$$

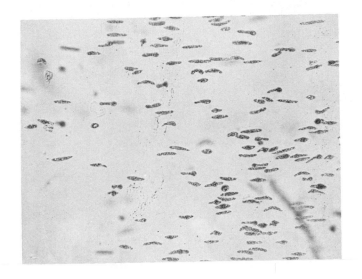

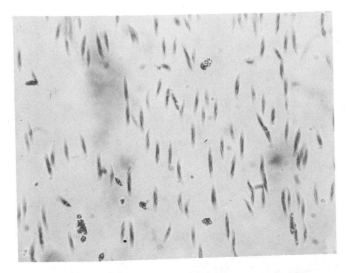

Fig. 4. Oriented Euglenae. The arrow gives the field direction.

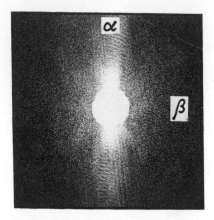

Fig. 5.   Light scattering pattern from oriented Euglenas. The
          arrows indicate the direction of the orienting field.
          α is the position of the photodetector chosen to reveal
          the intensity modulations of the scattered light;
          β is the photodetector position chosen to reveal the
          Doppler shift of the scattered light.

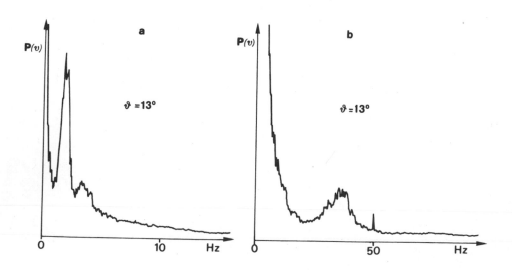

Fig. 6.   Homodyne spectra showing the light intensity modulations
          due to: a) Euglena body rotation, b) Euglena flagellar
          beating. After Ascoli et al. (1978a).

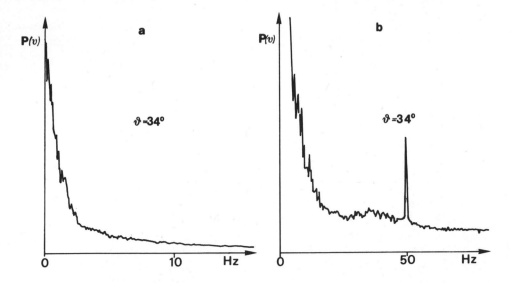

Fig. 7.   Geometry of light scattering in the experimental set up
          used to reveal light intensity modulations; $\vec{K}_o$ and $\vec{K}_s$ are
          the wavevectors of incident and scattered light respectively,
          and $\vec{V}$ is the cell velocity.

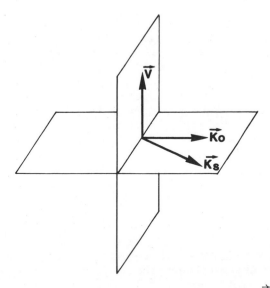

Fig. 8.   Homodyne spectra for Euglena moving with $\vec{V}$ in the
          scattering plane and perpendicular to the laser beam.
          Light intensity modulations are strongly reduced. After
          Ascoli et al (1978a).

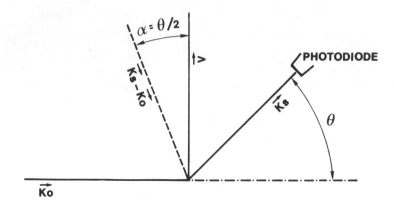

Fig.9.   Geometry of light scattering in the experimental set up
         used to reveal the Doppler shift of scattered light.

where v is the swimming speed, λ the laser wavelength, α the angle
between $\vec{v}$ and $\vec{K}_s\vec{K}_o$ and θ the scattering angle. Now, if oriented
Euglenae are of exactly the same size, they scatter the same light
intensity on the photodetector. Ascoli et al (3) have shown
that in this case the heterodyne spectrum is homologous with the
distribution of the oriented Euglenae swimming speeds, according
to relations (1). Thus the line in fig.10 shows the distribution
of swimming speeds; relation (1) is verified by changing the
scattering angle, as shown in fig.11. The mean speed of a
population of Euglenae depends on the growth conditions and we
find values in the 100-300µm/sec range.

    Orienting the samples is therefore a sueful technique in
measuring the motion parameters of Euglenae and in studying the
effects of external physical factors on them. Certainly,
orientation involves a marked change in the physico-chemical
properties of the cells' environment, but no damage to the cells
has been observed; as already observed by Wolken (1), the only
effect of applying a field is an increase in Euglena speed with
increasing field intensity. By comparing the effects of the
radiofrequency field with those of temperature, Ascoli et al (4)
have shown that the field does no more than warm up the sample
through dissipation dependent on the conductivity of the medium,
so inducing an increase in the motor activities of the oriented
cells. Therefore the quasi-elastic light scattering technique
can be used on oriented samples to investigate the action of
light on the motion of Euglenae.

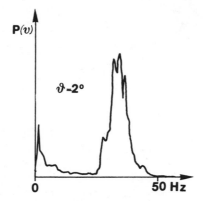

Fig. 10. Heterodyne line giving the swimming speed distribution
          of oriented Euglenae. After Ascoli et al. (1978a).

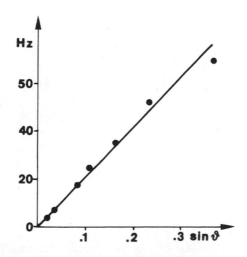

Fig. 11. Doppler shift due to the translation motion of Euglenae
          as a function of sin θ. After Ascoli et al (1978a).

Photokinesis Experiments

An action spectrum of photokinesis in Euglena gracilis
cultures has been reported by Wolken (1), but the mechanism
involved was unknown, even if Wolken hypothesized that the effect
might be controlled by the same pigment that controls Euglena
phototaxis.

Our experiments have not confirmed the hypothesis of Wolken
while we have found a photokinetic effect controlled by the
Euglena photosynthetic apparatus (5). Euglenae, which had been
grown in an autotrophic medium(where energy is supplied by the
photosynthetic apparatus) under constant illumination of 3000 lux,
were kept in the dark for many hours and then illuminated with red,
green and blue lights. The red and blue wavelengths used
6770 Å and 4310 Å, respectively) were near the maxima of the
chlorophyll (a) absorption spectrum, while the green light wavelength (5210 Å) corresponds to a low value of the chlorophyll (a)
absorption spectrum. The mean speed measurements obtained by the
laser light scattering technique on oriented Euglenae (fig.12) show
that only the red and blue lights increase the mean speed of cells,
while the green light and darkness decreased the Euglenae mean
speed. The effect is strongly dependent on the growth medium; in
fact, Euglenae grown in the dark or in an autotrophic medium (where
energy is supplied by the respiratory apparatus) are not
photokinetic, even if they are phototactic (6). We may conclude
that this effect is not related to the mechanism that causes
phototaxis but to the photosynthetic process.

Flagellar Beating Measurements in Unoriented Samples

The experimental approach thay proved effective for Euglena
gracilis cannot always be easily applied to other algae. In fact, many
microorganisms cannot be oriented by an electric field (spherical
microorganisms don't orient, sea-inhabiting microorganisms
cannot be resuspended in distilled water, etc.). Besides changes
in the physico-chemical properties of the medium and the time
required to prepare the samples are a source of difficulty in
many biological problems. On the other hand many biological
investigations only call for measurement of the motility of a
cell population. So we require a motion parameter that was a
good index of cell motility and that could be measured by a laser
light scattering technique even in unoriented samples. The only
easily measurable Euglena motion parameter in unoriented samples
is the flagellar beating frequency. Fig.13 shows the distribution
of flagellar beating frequencies obtained by direct detection of
the light scattered by unoriented Euglenae. We have also tested
the effectiveness of this parameter as an index of cell motor
activity by measuring its dependence on temperature in three
different experimental conditions and by comparing this dependence
with the dependence of speed on temperature in oriented samples –

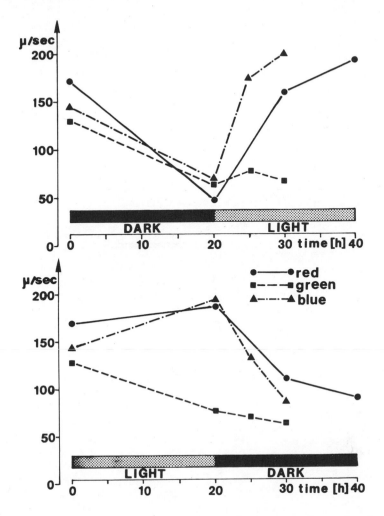

Fig. 12   Mean speed of oriented <u>Euglenae</u> versus irradiation time.
The irradiance is 2mW/cm for red light, 1mW/cm for
green light, and .5 mW/cm for blue light.

the only samples where such comparison is possible. The results
of fig.14 show that the dependence of flagellar beating on
temperature is not affected by changes in physico-chemical
conditions induced by orientation, and that the flagellar beating
frequency is proportional to the swimming speed when the sample
temperature is varied within a physiological range. In fact, a
deviation from proportionality is observed in the temperature
range in which the cells die.

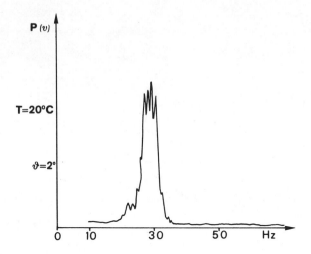

Fig. 13.   Flagellar beating frequency.   Distribution in unoriented
           Euglenae.

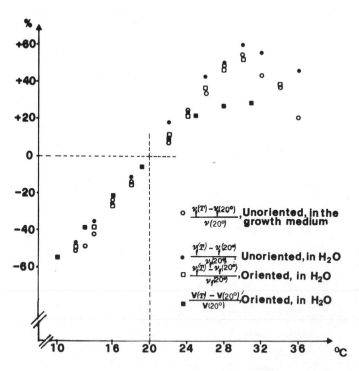

Fig. 14.   Dependence of Euglena flagellar beating and speed on
           temperature.

The flagellar beating frequency has been measured in many
microorganisms without orienting the sample. Spectra obtained
from samples of Haematococcus pluvialis, Dunaniella salina and
Ocromonas danica, which are flagellated algae of different shapes
and sizes which live in different environments, are shown in Fig.
15. The frequency of the maximum of the lines occurring in these
spectra is independent of the scattering angle (fig.16 and fig.17),
while the line width does not show the same dependence on θ in
these three algae.

Fig. 16 shows a line enlargement with increasing θ for
Haematococcus pluvialis, while the line width doesn't depend
on θ for Ocromonas danica (fig. 17) and Euglena gracilia. This
difference seems related to the light-scattering pattern, that is,
to the shape of microorganisms, Haematococcus pluvialis is quasi-

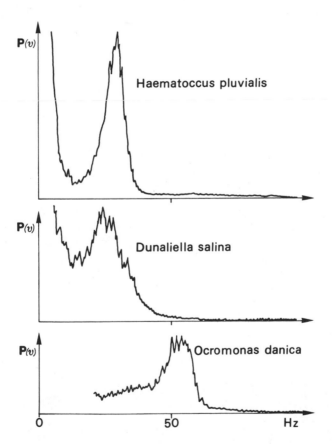

Fig. 15. Homodyne spectra giving the flagellar beating frequency
          distributions for three flagellated algae.

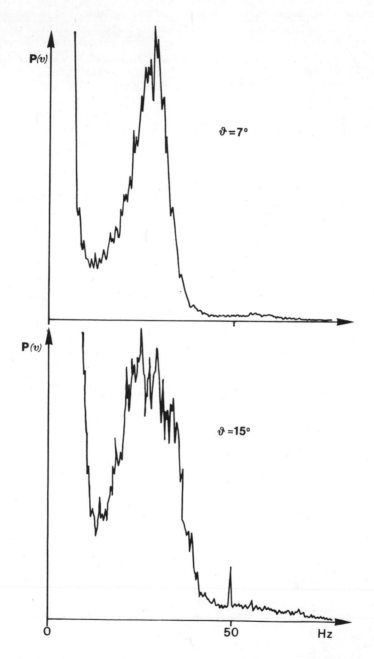

Fig. 16. Homodyne spectra at two values of θ for <u>Haematococcus</u> <u>pluvialis</u>.

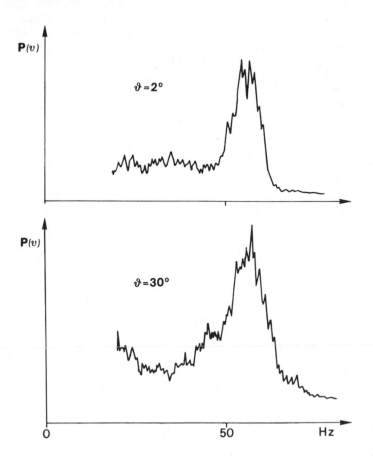

Fig. 17. Homodyne spectra at two values of θ for Ocromonas
          danica.

spherical microorganisms while Ocromonas danica and Euglena
gracilis are elongated cells. In any case,for all these algae the
flagellar beating frequency can be easily measured at small
scattering angles and this kind of measurement needs a simpler
and more rapid experimental procedure than the measurement of
the speed of oriented samples. It is, in fact, performed on cells
living in the growth medium without any manipulation and homodyne
detection is less critical than heterodyne detection. This kind
of measurement appears to be useful in studying biological
phenomena.

References

1. J. J. Wolken, "Euglena", Meredith Publishing Co., Des
    Moines, Iowa (1967).
2. J. L. Griffin and R. E. Stowell, Exp.Cell Res., 44:648 (1966).
3. C. Ascoli, M. Barbi, C. Frediani and A. Muré, Biophys.J.,
    24:585 (1978).
4. C. Ascoli, M. Barbi, C. Frediani, D. Petracchi, Biophys.J.,
    24:601 (1978).
5. C. Ascoli, C. Frediani and W. Nultsch, 1975 – A photokinetic
    effect in Euglena gracilis, P.116 5th International Biophysics
    Congress Copenhagen.
6. A.Checcucci, G.Colombetti, R. Ferrar, F. Lenci, Photochem.
    Photobiol., 23:51 (1976).

HYDRODYNAMIC AND CRITICAL PHENOMENA: RAYLEIGH-BENARD INSTABILITY
AS A CRITICAL PHENOMENON AND CRITICAL FLUCTUATIONS REDUCED BY
SHEAR, BOTH RELATED TO MEAN FIELD BEHAVIOUR

P. Bergé, D. Beysens, M. Dubois, J. Wesfreid
and M. Gbadamassi

SPSRM, CEM Saclay, B.P. No. 2
91190 GIF-sur-YVETTE
France

In the case of second order phase transitions, all the
critical behaviours are generally affected by the influence of the
fluctuations of the order parameter. The influence of fluctuations
can be evaluated through the well-known Ginsburg criterium. When
these fluctuations are important, the values of the critical
exponents significantly differ from that expected from mean-field
theory.

Nevertheless, we will show that particular systems related
to hydrodynamical flow allow the mean field behaviour to be
recovered. In particular, we present the results obtained in
two cases : i) hydrodynamical instabilities, where the characteristic
length is large compared to the scale of the thermodynamical
fluctuations and ii) a critical fluid submitted to a shear flow
which reduces and makes anisotropic the critical fluctuations.

RAYLEIGH-BENARD INSTABILITY

The Rayleigh Benard instability in a horizontal fluid layer
heated from below, is characterised at first by two points:- there
is a critical threshold i.e. the motion sets on for a given value of
$R = R_C$ (critical value of the Rayleigh number). - the motion is
spatially organised under the form of rolls, the wavelength of which
is $\Lambda \simeq 2d$; d is the depth of the fluid layer. At the onset, there is
a symmetry breaking, suggesting an analogy with a 2nd order phase
transition. In this case, the Fourier amplitude $\tilde{V}$ of the spatial
velocity mode can be taken as the order parameter (1). The
crucial fact is the following: The wavelength of the unique
Fourier component describing the spatial variation of the
order parameter is of macroscopic scale ($\simeq d$). This scale

defines then a macroscopic volume inside which thermodynamical fluctuations will be of negligible importance : we may then expect to have a mean-field behaviour.

The experimental results confirm this expectation.   Indeed, the measurements of the amplitude of the convective velocity showed the behaviour (2).

$$\tilde{V} = V_o \left(\frac{R - R_c}{R_c} \right)^\beta \text{ with } \beta = 0.5$$

Moreover, the analogous of the "correlation length" has been found in the Rayleigh Benard instability, as well in the supercritical regime (3) as in the subcritical one (4).

When $R > R_c$, the "influence" length ξ_+, obtained from the spatial modulation of the velocity amplitude near a lateral boundary (where $V = 0$), follows the relation:

$$\xi_+ = \xi_o \left(\frac{R - R_c}{R_c} \right)^{-0.5}$$

when $R < R_c$, the influence length ξ_- determines, for example, the penetration length of rolls, induced by an external local flow, and

$$\xi_- = \frac{\xi_o}{\sqrt{2}} \left| \frac{R - R_c}{R_c} \right|^{-0.5}$$

In the two cases, $\nu = 0.5$ and $\xi_o = 0,54d$, a universal value.

Lastly the study of the critical slowing down has clearly established the variation of the characteristic time τ : (4) (5).

$$\tau = \tau_o \left(\frac{R - R_c}{R_c} \right)^{-1}$$

As one can see, all the critical exponents are classical, as expected from a mean-field Landau theory.

Critical Fluid Under Shear

By contrast with the case of the Rayleigh-Benard instability, in the case of a critical fluid ξ_o is a mean intermolecular distance ($\sim 2\text{Å}$). Fluctuations are important and the critical exponents are different from the mean-field exponents. Specially the susceptibility χ varies near the critical temperature T_c as

$$\left(\frac{T - T_c}{T_c}\right)^{-\gamma}$$ with T the temperature and $\gamma \simeq 1.24$ {6}.

But when such a fluid, with critical fluctuations of lifetime τ, is submitted to a shear flow of rate S (lifetime S^{-1}), it is intuitively obvious that an effect should be visible only if the fluctuations have enough time to "feel" the shear, i.e. only in the region $S\tau > 1$. The shear would make the fluctuation anisotropic with respect to the flow direction, and it is understandable that a reduction of the fluctuation size would occur along the shear direction. This means that the temperature at which the fluctuations diverge (critical temperature T_c) is lowered by the shear :
$T_c = T_c(S) < T_c(0)$.

These points have been verified both theoretically (7) and experimentally (8-10). Moreover, the shear-induced anisotropy makes negligible the influence of fluctuations, as in uniaxial dipolar-coupled ferromagnets (11).

A cross-over in the susceptibility is predicted (8) and experimentally checked (9) by measuring the scattered intensity at small wavenumbers as a function of the shear rate S. The striking feature is the following : as far as $S\tau < 1$ the usual $\gamma = 1.24$ exponent is measured; on the contrary when $S\tau > 1$ the mean-field $\gamma = 1$ exponent is found, obviously for the first time in the case of a critical fluid!

References

1. P. Bergé, Fluctuations, Instabilities and Phase Transitions. Nato Advanced Study Institute, Plenum Press, New York, Vol. B.11, 323 (1975).
2. M. Dubois and P. Bergé, J. Fluid Mech., 85:641 (1978).
3. J. Wesfreid, Y. Pomeau, M. Dubois, C. Normand and P. Bergé, J.Physique (paris), 39:725 (1978).
4. J. Wesfreid, P. Bergé and M. Dubois, Phys. Rev., A.19:1231 (1979).
5. R. P. Behringer and G. Ahlers, Phys. Letters, 62 A:329 (1977).
6. J. G. Le Guillou and J. Zinn-Justin, Phys. Rev. Lett., 39:95 (1977).
7. A. Onuki and K. Kawasaki; a) Prog. Theor. Phys. Supplement 64:436 (1978)   b) Phys. Letters, A72:233 (1979),   c) to be published in Ann. Phys. (N-Y).
8. D. Beysens, M. Gbadamassi and L. Boyer, Saclay preprint Dph-G/PSRM/1568 (1978) to be published in Phys. Rev, Lett.

9. D. Beysens and M. Gbadamassi, Saclay preprint Dph-G/PSRM/1591
    (1979) to be published in J. de Phys. Lett.

10. D. Beysens Saclay preprint Dph-G/PSRM/1611 (1979), to be
    published in "Ordering in Strongly Fluctuating Condensed
    Matter Systems", T. Riste, ed., Plenum Press, London.

11. J. Als-Nielsen, Phys. Rev. Lett., 37:1161 (1976).

INTERFACE FLUCTUATIONS OF GROWING ICE CRYSTALS

J. H. Bilgram and P. Böni

Laboratory of Solid State Physics
Swiss Federal Institute of Technology
Zürich, Switzerland

ABSTRACT

Light scattering experiments at the surface of a growing
ice single crystal show that fluctuations at the solid liquid
interface occur once a minimum growthrate has been exceeded. This
quasi elastic scattering vanishes again after the surface has
been brought to equilibrium conditions. Scattering intensity
and line width have been measured as a function of scattering
angle and angle of incidence of the laser beam on the interfacial
plane. The results indicate that far from equilibrium a phase
boundary layer is nucleated. Its thickness is of the order of
magnitude of wavelengths of the laser light.

Three different models for the scattering process are
discussed.

INTRODUCTION

This experiment has been started because on a microscopic
scale little experimental information is available about the
dynamics of the freezing process. Well studied are the equilibrium
properties of crystals very close to the melting point and the
properties of supercooled liquids. We study the dynamics of the
freezing process at the solid liquid interface at conditions far
from equilibrium (1,2). Light scattering is used because this
technique is potentially able to provide information on dynamics
of small sampling volumes. We use the system ice-water because
of its low absorption of visible light and the availability of
highly perfect single crystals in our laboratory.

EXPERIMENTAL

     High purity water is prepared by zone refining. In order to
exclude any dust the experiment is performed in situ during
zone refining. The experimental setup is shown in figure 1.
This zone refining apparatus is placed in a coldroom at -18°C.

     During the experiment the ice-crystal is lowered into the
cooling bath exactly at the growthrate,so that the ice-water
interface does not move relative to the optical system. For
experimental details see (2). Ψ_0 is the angle of incidence of
the laser beam. The photomultiplier can be rotated around the
growth tube. It detects light scattered parallel to the interface
plane. θ is the angle between the projection of the incident
laser beam on the interface plane and the direction of the observed
scattering. For $\Psi_0 = 90^\circ$, θ is the scattering angle.

THE INTENSITY OF THE SCATTERED LIGHT

a)   The Intensity Hysteresis

     An ice single crystal which is slowly growing into water can
show a facet or a rough interface. The facet can develop if
the interface is a {0001} plane, whereas for other orientations
the interface is rough. We have measured the intensity of light
quasielastically scattered from both types of ice-water interfaces
as a function of the growth rate v. We have found in both cases
a hysteresis behaviour in this dependence (fig.2). There are two
stationary regimes where single crystal growth can occur:

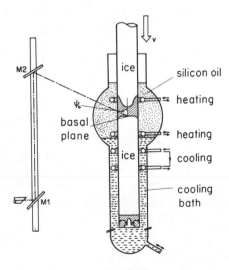

Fig. 1.   Zone melting apparatus.

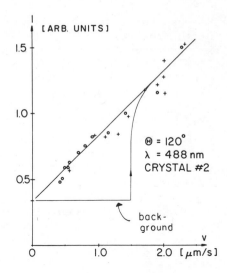

Fig. 2. Hysteresis behaviour of the scattered intensity as a
         function of growth rate.

     close to equilibrium (state 1)
     far from equilibrium (state 2).

Starting with a crystal at equilibrium conditions (state 1) no
light is scattered quasielastically from the interface. Only
unmodulated straylight (background) is detected. When the growth
velocity exceeds 1.5 μm/s the onset of quasielastic light
scattering occurs. The intensity is proportional to the growth
rate as long as it is positive. The maximum growth rate achieved
was 5 μm/s. After reverting to equilibrium conditions (state 1)
quasielastic scattering vanishes and can be initiated again by
exceeding the growth rate of 1.5 μm/s. This behaviour is observed
at the facet as well as at rough interfaces. Dynamical scattering
occurs in a thin interface region only. In liquid water at
temperatures near 4°C no unshifted light is scattered at all.   In
our crystals intensity of the Raman scattering light is large
compared to the intensity of the Rayleigh and Brillouin components
together.

     From the angular dependence of the intensity of the scattered
light one concludes that the upper limit of the diameter of the
scattering inhomogeneities is below the wavelength of light.

b)   The Scattering Region

     For a determination of the nature of the scatterers it is
necessary to know the properties of the scattering region in three

dimensions. We will discuss three different models that give different dependence of the scattered intensity on the angle of incidence for a constant density of illumination. In all models the critical angle of total reflection ψ_{crit} plays an important role.

Model A) Particles e.g. ice clusters are suspended in water in front of a sharp plane interface. For $\psi_o < \psi_{crit}$ the presence of the phase boundary has little influence on the scattered intensity. For all $\psi_o > \psi_{crit}$ the primary laser beam is totally reflected at the phase boundary and passes twice the sampling volume thus doubling the illumination density. Hence with increasing ψ_o the scattered intensity doubles at ψ_{crit}.

Model B) Light is scattered by a sharp but rough interface. The intensity scattered in any direction from a rough surface has been calculated by Mandelstam (3) as a function of the angle of incidence of the illuminating beam.

The following calculations are based on the extension of this work by Andronow and Leontowitz (4). These authors solved Maxwell's equation for a sinusoidally corrugated interface under the assumption that the corrugation period $\Lambda = 2\pi/p$ is large compared to its amplitude ζ_p. The square of the amplitude f_ψ of the scattered light is given by

$$f_\psi^2 = 4K_o^2 |\overline{\zeta_p}|^2 |F|^2 \frac{(tg\psi\sin\psi_o - ((n^2\sin^2\psi_o - 1)(n^2\sin^2\psi - 1))^{1/2}\frac{\cos\alpha^2}{\cos\psi}}{(n^2 tg^2\psi_o - 1)(n^2 tg^2\psi - 1)}$$

where $K_o = \frac{2\pi}{\lambda_o}$ is the wavenumber of the laser light in vacuum and F its amplitude. ψ and α define the polar and azimutal angle of the scattered light. The angle θ in our measurements is defined as $\theta = 180^o - \alpha$. n is the relative index of refraction

$$n = \frac{n_{water}}{n_{ice}} = \frac{1.338}{1.3135} .$$

The square of the scattered field is plotted in figure 3 versus the angle of incidence ψ_o for $\psi = 85^o$ and $\theta = 90^o$. There is one maximum at the critical angle of total reflection ψ_{crit} with an intensity that is four times that for small angles. This maximum is due to the fact that the evanescent wave has a maximum amplitude at this angle thus producing an electrical field strength at the interface of twice that of the incident beam (5).

Model C) There is an intermediate layer formed between water and ice. Light scattering occurs at inhomogeneities in this layer. This is a complex model because it contains unknown

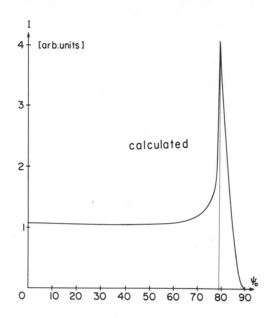

Fig. 3. Calculated scattered intensity as a function of the
angle of incidence ψ_o for $\psi = 85^o$ and $\theta = 90^o$.

parameters, in the simplest case a thickness d and a refractive
index n. In the limits of d→∞ or d→o this model approaches the
model A or B, respectively. For a thickness of at least the order
of magnitude of λ_{laser} the layer will form an optical waveguide.
It will be leaky except for special values of n and d.

c)   Intensity Measurements

We have measured for $\psi \approx 85^o$ and $\theta = 90^o$ the scattered
intensity as a function of ψ_o. The results are plotted in figure
4. If ψ_o approaches ψ_{crit} we find an increase of the scattered
intensity about by a factor four. This rules out model A. The
amount of this increase suggests that light from the evanescent
wave might be scattered and that the scattering region has a
thickness of a few wavelengths of light or less. For $\psi_o > \psi_{crit}$ we
do not find the decrease of the scattered intensity as expected
for a rough interface according to model B. Instead we find a
second maximum. The height of this maximum differs from run to
run but it is always present. The wave travelling along the
interface on the ice side appears to have a stronger field than
would be expected for a statistically rough surface. The observed
dependence of the scattered intensity on ψ_o can be interpreted as
resulting from a leaky wave guide.

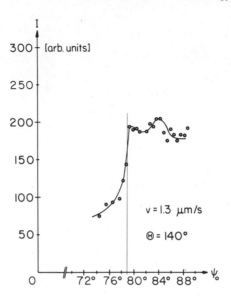

Fig. 4.   Measured scattered intensity versus ψ_0 for $\psi \simeq 85°$ and
$\theta = 90°$.

THE LINE WIDTH OF THE SCATTERED LIGHT

     Measurements of the line width of the quasielastically
scattered light can give two kinds of information: ·

i)   From the correlation function and its dependence on the
       scattering vector one can deduce the number of different
       decay mechanisms and their characteristic times.

ii)   Comparison of measurements with the scattering vector in
       the interface plane with measurements where the scattering
       vector has a component perpendicular to this plane yields
       information about the isotropy of the decay process.
       (e.g. it is possible to distinguish volume and surface
       diffusion). The models A, B, C correspond to different
       decay mechanisms.

The Scattering Region

Model A

     Dispersed particles undergo Brownian motion. The line width
measurements then yield a hydrodynamic radius. It is improbable
that all clusters have the same diameter so that a distribution
of decay times will be expected.

Model B

The average interface shall be the xy-plane and the rough interface shall be given by a function $\zeta(x,y)$ which can be represented as a two-dimensional fourier integral in terms of the wave vectors $\vec{p} = \binom{px}{py}$. If the scattering vector $\vec{K}$ is parallel to the interface the scattered light can be attributed to the fourier component $\vec{p} = \vec{K}$. If the scattering vector has a component perpendicular to the interface the scattered light can be attributed to the fourier-component corresponding to the projection of the scattering vector on the interface.

The thermodynamical assumption is the following: The growth rate v of the ice crystal is proportional to the deviation of surface temperature from its equilibrium value

$$v = \mu \cdot \Delta T.$$

Under these assumption it is for the range of the fourier components under investigation only the interface curvature which determines the dynamics. Based on the Gibbs – Thomson equation one obtains for the decay time of a fourier component with the wave vector p (1,2):

$$\frac{1}{\tau} = \Gamma = \frac{\mu \; \sigma_{s\ell} \cdot p^2}{\Delta S_f}$$

$\sigma_{s\ell}$ designates the solid – liquid interfacial energy and ΔS_f the melting entropy. The proportionality μ has been measured by Hillig (5). The microscopic physics of the observed relaxation process is hidden in this phenomenological constant. As long as the scattering plane is identical with the interface plane as it was in the earlier experiments one has $K^2 = p^2$. However for experiments where the scattering vector has a component perpendicular to the interface a deviation of Γ from $\Gamma \sim K^2$ has to be expected.

Model C

It is assumed that a layer of an intermediate phase is nucleated when the system is far away from equilibrium. In this layer the molecules are not all linked by hydrogen bonds as in the ice lattice. Structural fluctuations modulate the light. This process is similar to that which gives rise to light scattering in glasses (7). The measured correlation time is characteristic for the structural relaxations. The existence of such a layer has yet to be proved by other experiments e.g. by Brillouin spectroscopy.

Line Width Measurement

    Measurements of the correlation function of the photoncounts
can be fitted by one single exponential. There is no indication
for a distribution of decay times. This is an argument to rule
out model A. For experiments with the scattering vector in the
interface plane the line width increases with the square of the
scattering vector. The line width does not depend on growth
rate. For crystals where the growth direction is perpendicular
to the c-axis (rough interface) the decay time is about twice
that for the {0001}. It is possible to interpret the measured
line width in terms of model B (2).

    Additional measurements where we do no longer illuminate
the interface in grazing incidence but choose $\psi_o = 65^o$ are
plotted in figure 5. Although the scattering vector has a component
perpendicular to the interface we do not find a deviation from

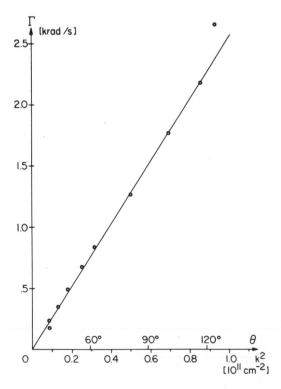

Fig. 5.   Line width of the scattered light as a function of the
          square of the scattering vector.

          The angle of incidence is $\psi_o = 65^o$.

$\Gamma \sim K^2$. This observation is compatible with a three dimensional
scattering system with an isotropic characteristic time. The
proportionality between Γ and K^2 varies from crystal to crystal
by about ±15% (only {0001} crystals were investigated so far).
The isotropy of the characteristic time was confirmed by line
width measurements with fixed K^2 and variable direction of
K (fig. 6). (Variation of ψ_o keeping $\psi \approx 90^o$ and $\theta = 90^o$).
As shown in figure 6 there is no influence of the orientation of
the scattering vector on line width. It is only the magnitude
that determines the line width.

CONCLUSION

     In table 1 the predictions of the three interface models are
compared with the experimental results. The scattering intensities
for $\psi_o > \psi$ and the isotropy of Γ seem to be the most sensitive
properties for distinguishing these models. For model C gives
the best agreement with the experiment.

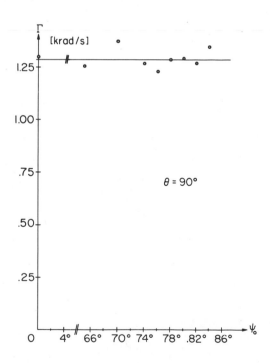

Fig. 6.   Line width of the scattered light for various angles
          of incidence but constant magnitude of the scattering
          vector.

TABLE 1

Comparison of the interface models and the experimental results

Model	Modulated Intensity		$\Gamma(\vec{K})$		Number of relaxation times
	$\dfrac{J(\psi_{crit})}{J(o)}$	$J(\psi_o > \psi_{crit})$	$\vec{K}$ in the interface plane	$\vec{K}$ not in the interface plane	
A diffusing clusters	2	const	$\sim K^2$	$\sim (K_x^2 + K_y^2 + K_z^2)$	many
B rough surface	4	$\rightarrow o$	$\sim K^2$	$\sim (K_x^2 + K_y^2)$	1
C Intermediate layer	dependent on thickness	more than one maximum possible	$\sim K^2$	$\sim (K_x^2 + K_y^2 + K_z^2)$	?
Experiment	4	two maxima no decrease	$\sim K^2$	$\sim (K_x^2 + K_y^2 + K_z^2)$	1

ACKNOWLEDGEMENTS

The authors thank W. Känzig for his steady interest and many fruitful discussions and careful reading of the manuscript.

This work is supported by the Swiss National Science Foundation.

REFERENCES

1. J. H. Bilgram, H. Güttinger and W. Känzig, Phys.Rev.Letters, 40:1394 (1978).
2. H. Güttinger, J. H. Bilgram and W. Känzig, J.Phys.Chem.Solids, 40:55 (1979).
3. L. Mandelstam, Ann.Physik., 41:609 (1913).
4. A. A. Andronov and M. A. Leontovicz, Z.Physik., 38:485 (1926).
5. O. Bryngdahl, in "Progress in Optics XI", E. Wolf, ed., North-Holland, Amsterdam (1973), p.167.
6. W. B. Hillig, in "Growth and Perfection of Crystals" edited by R. H. Doremus, B. W. Roberts and D. Turnbull, John Wiley, New York (1958), p.350.
7. J. Schroeder, in "Treatise on Materials Science and Technology" Vol.12, M. Tomozawa and R. H. Doremus, ed., Academic Press, New York (1977), p.157.

BUOYANCY DRIVEN INSTABILITIES IN TWO COMPONENT SYSTEMS

M. Giglio, S. Musazzi, U. Perini and A. Vendramini

CISE S.p.a.
P.O.B. 3986, 20100 Milano
Italy

INTRODUCTION

In this communication we will describe some experimental results on buoyancy driven instabilities in two component liquid systems confined between two horizontal conducting plates.

The configuration is similar to that of the well known Rayleigh-Benard Instability (RBI). At variance with the RBI case, however, the density changes in the fluid are caused not only by temperature variations, but also by concentration variations. A fundamental role is played here by the so called Soret effect (1), which couples the mass flow to the temperature gradient. As a consequence of this effect, when a temperature gradient is applied to a binary system, one observes at steady state a concentration gradient grad $c = - s_T c(1-c)$ grad T where s_T is the Soret coefficient, The magnitude and sign of this coefficient depend on the type of components and on the concentration. It must be noticed that if, as customary, the solute concentration c is that of the heavy component, a positive sign for s_T implies that the solute migrates to the cold plate. The reverse is true for a negative s_T. With regard to the problem of the stability of a two component system, four cases must be considered, according to the sign of the temperature gradient and that of the Soret generated concentration gradient. Temperature gradients and concentration gradients may have either a stabilizing or destabilizing effect. One case is of no interest, that is when both gradients are stabilizing.

In this communication we will present results regarding the other three cases. Two cases are very similar in behaviour.

215

These are the cases characterized by a destabilizing action of the
concentration gradient. Here the density variations imposed by
the thermal gradient via the thermal expansivity of the system
play essentially no role and it is only the adverse solute
concentration gradient which is responsible for the instability
Data on these cases are drawn from published work (2) or are a
limited and preliminary version of data to be presented in an
extended work under way of completion (3). The third case is
that characterized by a stabilizing concentration gradient and by
a destabilizing temperature gradient. While the first two cases
have strong similarities with the RBI case, the last one is
definitely more complex and we are just beginning to investigate
on it. We will present here some of the preliminary results
obtained so far.

STABILITY DIAGRAMS AND CHOICE OF SAMPLES

     Stability diagrams have been obtained by a number of authors
(4-6) and we will simply comment on some aspects which are related
to the choice of the samples we have used. Let us first consider
the two cases in which the solute concentration gradient is
stabilizing (counting clockwise, see quadrants II and IV in Fig.1.

     In Fig. 1 the symbol R is the Rayleigh number

$$R = \frac{g \alpha \Delta T a^3}{\chi \eta} \tag{1}$$

Here g is the acceleration due to gravity, $\alpha = -(1/\rho) (\partial\rho/\partial T)$ is
the coefficient of expansion, a the fluid thickness, χ the
thermal diffusivity and η the viscosity. A positive value for
R means heating from below. The parameter S in Fig.1 relates to
the magnitude of the thermally induced concentration gradient

$$S = \frac{\beta s_T c(1-c)}{\alpha} \tag{2}$$

where $\beta = (1/\rho) (\partial\rho/\partial c)$. That is, S in the ratio between density
changes due to thermally induced concentration gradients and
density changes due to thermal expansion. As indicated in Fig.1,
for large (positive or negative) values of S, the threshold
condition is given by

$$R = R (\frac{\chi}{D}) S = 720 \tag{3}$$

where D is the mass diffusion coefficient. Furthermore, the
stability analysis indicate that the wavevector k_c of the mode

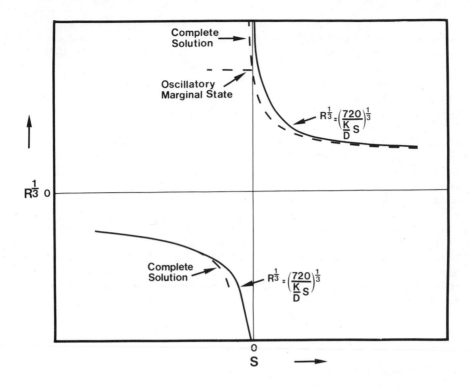

Fig. 1. Stability diagram for two component systems in a
         temperature gradient (Taken from Ref.5).

excited at threshold should be k_c = o. In practice the actual
critical wavevector is determined by the lateral confinement of the
sample. Recalling that for the RBI the threshold for rigid
conducting boundaries is R = 1708, one then expects that for
mixtures the critical temperature difference should be substantially
lower, at least for cases characterized by large values of (χ/D)
and S. Macromolecular solutions in general satisfy both these
requirements. For our studies we have chosen polymer solutions
for which data on S and D were available. The first study (2)
was on a 0,7% by weight solution of polyvinylalcohol (PVA) in
water. The sample was polydisperse and we did not have estimate
for its average molecular weight. The z-average diffusion
coefficient however was D=1.35 x 10^{-7} cm^2/sec, a fairly small
value which obliged us to use a thin cell, 0.5 mm high, in order
to keep time constants within reason.

This system is characterized by a negative value for s_T and
therefore was heated from above (quadrant IV).

The second study, whose results are yet unpublished, was on

a 1% methyl ethyl ketone solution of monodisperse polystyrene molecular weight 4000. This sample has a positive s_T and heating was provided from below (quadrant II). A 1 mm high sample was used in this case.

Finally in quadrant I in Fig.1 we have the case of a stabilizing concentration gradient and a destabilizing temperature gradient (heating is done from below and s_T is negative). For negative values of S one expects that the threshold for the convective instability should be somewhat higher than the RBI, due to the presence of a stabilizing concentration gradient. As one can notice in Fig.1, as ΔT, and therefore R is made larger, one encounters a line of overstability. That is, linear stability analysis predicts that at threshold the system will abandon its quiescent state via an oscillatory behaviour of the relevant variables, at variance with what happens along the ordinary stability lines in quadrants II and IV. The critical wavevector in this case is, at least for small values of S, very close to the RBI value $k_c = (2\pi/\Lambda) \simeq (2\pi/2.02a)$. The system we have chosen for this study is a water ethanol mixture, ethanol 10% by weight). The sample was 5 mm high.

EXPERIMENTAL SETUP AND TECHNIQUE

We will briefly describe the experimental setup and the technique used for all the three cases.

The quantity that we have studied is the time evolution of the concentration gradient evaluated at midheight of the fluid sample. The technique used is a laser beam deflection technique (7). This technique is based on the fact that a light beam on traversing a fluid of length ℓ, with a refractive index gradient (dn/dz) orthogonal to the direction of propagation of the beam, undergoes a deflection

$$\Delta\theta = \ell \frac{dn}{dz} \tag{4}$$

where $\Delta\theta$ is measured in air. Of course (dn/dz) is the average value along the length of the beam inside the fluid. Quite generally the refractive index gradient (dn/dz) is related to both the temperature and concentration gradient

$$\frac{dn}{dz} = (\frac{\partial n}{\partial T}) \frac{dT}{dz} + (\frac{\partial n}{\partial c}) \frac{dc}{dz} \tag{5}$$

The second term is the interesting one, and it can be readily separated from the first since usually it evolves much more slowly. Refractive index gradients are determined using a beam from a low power He-Ne laser midly focused in the center of the

cell, and a beam scanning assembly which measures vertical beam
displacement at a given distance from the cell.

The cell consists of two aluminum alloy plates and a
cylindrical glass window. The two plates are of square cross
section, are identical in shape and have cylindrical expansions
which slide into the glass window which has a 40 mm OD, and is
10 mm high. The height of the cylindrical expansions is chosen
as to bring the thickness of the fluid slab to the desired value.

The temperature difference between the plates is controlled
by a temperature controller utilizing two Peltier heat pumps.
The error signal is provided by two thermistors placed in a d.c.
bridge, and located between the heat pumps and the cell plates.
The stability of the temperature difference is better than one
millidegree.

Measurements are performed in the following way. The
temperature difference between the plates is initially set equal
to zero and the system is made perfectly homogeneous. At time
t = 0 the temperature difference is switched on by changing the
resistance of the decade box inserted in one of the arms of the
sensing bridge. The angular deflection of the transmitted laser
beam is then monitored as a function of time.

RESULTS IN QUADRANTS II AND IV

A typical set of data obtained on the system PVA water (2) is
shown in Fig. 2. The first trace refers to a stable case.
The gradient builds up and remains constant as a function of time.
From the steady state value one derives s_T while the risetime
constant gives the diffusion coefficient D. Indeed one can show
(8) that this time constant is given by $\tau = (a^2/\pi^2 D)$, where a is
the fluid thickness. This type of stable behaviour is observed
when the temperature difference is small. As the temperature
difference is made larger one observes traces like those shown
in Fig.2b and 2c. In Fig.2b although grad c attains at first the
steady-state value that one would expect in the absence of the
instability (dashed line), after 4 h its value slowly decreases
and eventually attains a new, true steady-state value. Similar
behaviour can be observed in Fig.2c. Here grad c again attains
the false steady state that one would calculate on the basis of the
known value of s_T, but it leaves that value much earlier than
before, and the true steady state is markedly lower. A limiting
case is shown in Fig.2d. In this case grad c does not quite
make it to reach the dashed line and, after reaching a maximum,
rapidly decreases to the true steady state, exhibiting a heavily
damped oscillatory behaviour. Close to threshold, plots similar
to those shown in Fig.2 have also been obtained for the polystyrene
sample and we will not duplicate them here in a Figure. Examples

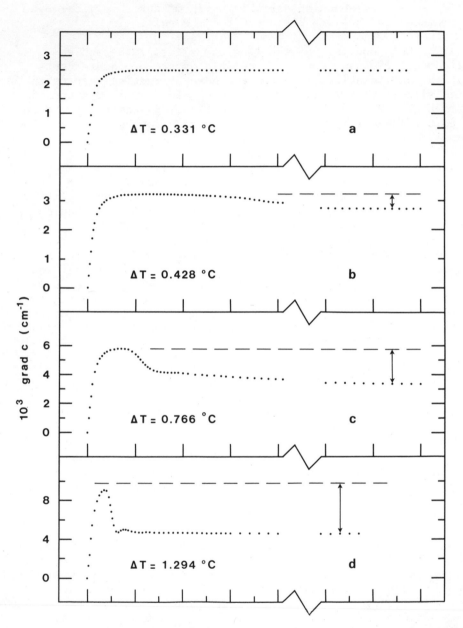

Fig. 2. Time evolutions for the concentration gradient for various
         ΔT_s. Polyvinylalcohol-water sample (Taken from Ref.2).

of traces relative to the polystyrene sample obtained in a region
fairly above threshold, where data on PVA are scant, are however
presented in Fig.3.

Let us now interpret qualitatively the curves in Fig.2. If
the temperature gradient gets large enough, convective motion
will set in inside the sample and this will give rise to a
periodic space structure, characterized by a critical wavevector
k_c. Considering for simplicity a rectangular geometry, the
concentration mode and the vertical component of the velocity
mode will have the form

$$\delta c = C(\varepsilon)\ w(z)\ \cos k_c x \tag{6}$$

$$v_z = B(\varepsilon)\ v(z)\ \cos k_c x$$

where $C(\varepsilon)$ and $B(\varepsilon)$ describe the steady state amplitude of the
modes as a function of the reduced distance from threshold $\varepsilon =$
$(\Delta T - \Delta T_c)/\Delta T_c$. The height dependence is described by $w(z)$ and
$v(z)$ which are assumed to be ε independent and have the appropriate
form to satisfy boundary conditions. As a consequence of the
concentration and velocity mode, we will have convected mass
flow across any horizontal plane.

$$J_{conv}(t,z) = \frac{1}{2} B\ (\varepsilon)\ C(\varepsilon)\ w(z)\ v(z) \tag{7}$$

Since at steady state the total mass flow J is zero,

$$J = \rho D \left(grad\ c\ + S_T(1-c)\ grad\ T \right) + J_{conv} = 0 \tag{8}$$

the steady state concentration gradient will be

$$grad\ c\ =\ -\ s_T\ c(1-c)\ grad\ T + \frac{J_{conv}}{\rho D} \tag{9}$$

The steady-state concentration gradient is then reduced by the
convective motion, a situation similar to that encountered with
the RBI where the convected heat flow partially destroys the
temperature gradient in the cneter of the fluid slab. Taking for
z=0 the position of the midplane, since s_T and D are known, the
actual value of J_{conv} can be derived from the measurements of the
true steady state concentration gradient.

The first quantitative result that can be obtained from
Fig.2 is then the behaviour of J_{conv} as a function of the distance
from threshold. This result is shown in Fig.4, which refers to the
PVA case. An alalogue plot can be derived for the polystyrene
sample. As one can see, the convected mass flow increases linearly
as a function of $\Delta T - \Delta T_c$.

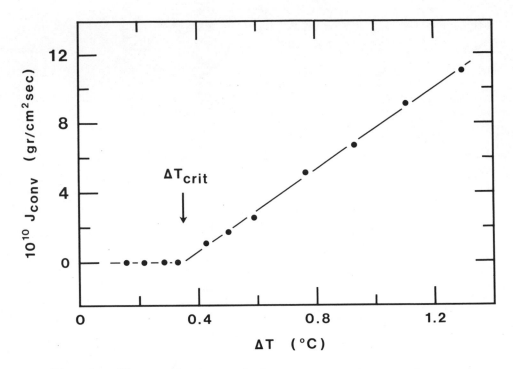

Fig. 3. Time evolutions of the concentration gradient for
various ΔTs Polystyrene sample.

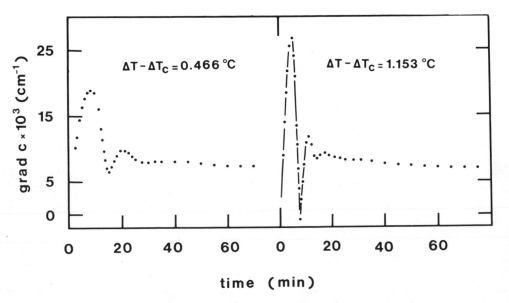

Fig. 4. Convected mass flow as a function of $\Delta T - \Delta T_c$
(Polyvinylalcohol sample).

We will comment on this result, as well as on the other experimental results, in the section dedicated to the comparison with the theory.

From the curve in Fig.2 one can also extract information on some dynamic properties of the system. One can notice that the above threshold the decay to steady state is characterized by time constants which are long close to threshold and become progressively shorter as one moves to higher temperature differences (let us ignore for a moment the oscillatory behaviour). The data on eht PVA case were not good enough to allow a quanitative estimate of these time constants. On the polystyrene sample however better traces could be obtained and these time constants could then be determined. The results are shown in Fig.5. The solid line through the points of the type $1/(\Delta T - \Delta T_c)$. Rough estimates from the PVA system are consistent with this result. This indicates that close to threshold the modes exhibit the phenomenon of the "critical slowing down" also encountered in classical thermodynamic second order phase transitions, and the dependence of the relaxation time as a function of the reduced distance ε from the critical point is the same.

Finally, fairly above threshold, damped oscillations appear. Again, this situation was not very well studied for the PVA case, and better traces were obtained with the polystyrene sample. At least one feature was very evident in these traces (see Fig.3). Although the damping is very strong in all cases, one can safely say that the period of the oscillations becomes shorter for larger ΔT_s. Only two cases are reported in Fig.3.

Let us now come to the comparison of all these experimental results with the theoretical predictions.

We will briefly describe now some of the theoretical results available in the literature and pertinent to our measurements. As a guideline we will refer to a recent single mode model proposed by Degiorgio (9) which accounts in a unified way for all the results described so far. Reference to partial results obtained in separate works will also be given. One assumes the following expressions for spatial distribution of the vertical velocity component and concentration

$$v_z = B(t)\, v(z)\, \cos k_c x$$
$$c = c_o + A'\,(z,t) + C(t)\, w(z)\, \cos k_c x$$

$$(10)$$

The quantity $A'(z,t)$ can be rewritten as

$$A'(z,t) = s_T\, c(I-c)\, \frac{\Delta T}{\alpha}\, z - A(t)\, f'(z) \tag{11}$$

In view of the small variations of c, the term $s_T\, c(I-c)$ is assumed to be constant. The dynamic equations are then three equations for B, C and Δ, where $\Delta = \dfrac{\partial}{\partial z}\, A'(z,t)\Big|_{z=0}$

$$\Delta = s_T\, c(I-c)\, \frac{\Delta T}{\alpha} - A(t)\, \left(\frac{df(z)}{dz}\right)_{z=0} \tag{12}$$

Notice here that Δ is precisely the quantity which is determined experimentally, the concentration gradient at midheight. It is made by a constant part, due to the undisturbed Soret effect, diminished by $A(t)\ (\frac{df}{dz})_{z=0}$

Without entering into the details of the calculations, the steady state solutions obtained by setting $\dot{B}=\dot{C}=\dot{\Delta}=0$ yield for the amplitude of the velocity mode and concentration mode solutions of the type

$$\begin{aligned} C_s(\varepsilon) &\propto \varepsilon^{1/2} \\ B_s(\varepsilon) &\propto \varepsilon^{1/2} \end{aligned} \tag{13}$$

Recalling that $J_{conv} = 1/2\, B_s(\varepsilon)\, C_s(E)$, we find that the experimental observation that the convected mass flow depends linearly on t is in agreement with the theoretical predictions. Analogour results can be found from perturbation analysis (5). The single mode model and the perturbation calculations produce however different results for the proportionality constant to be put into Eqs.13.

For the moment we cannot verify which theory gives more realistic estimates, but at least the dependence of J_{conv} on ε seems to be predicted firmly and confirmed experimentally. The complete sets of equations can generate sets of simplified equations, valid in limited regions of ε, by means of ordered adiabatic eliminations. Close to threshold one sets $\dot{B} = \dot{\Delta} = 0$ and obtains a unique equation for $C(t)$

$$\dot{C} = g\, D\, k_c^{\,2}\, E\, C - h\, \frac{D k_c^{\,4}}{\Delta_c^{\,2}}\, C^3 \tag{14}$$

where g and h are two constants.

This is a Van der Pol equation which predicts that the decay to steady state should be characterized by time constants of the type

$$\tau^{-1} = (g \; Dk_c^2) \; \varepsilon \tag{15}$$

in agreement with the ε dependence we have determined experimentally (Fig. 5).

Finally, fairly above threshold, two equations are necessary to describe the behaviour of the system. We will not report here the system of equations, but we will simply say that computer solutions reproduce qualitatively in a satisfactory way the observed damped oscillatory behaviour (9). Furthermore, linearization of these two equations around steady state produce the result that oscillation will appear only above a well defined threshold and that the frequency should increase as ΔT is increased. Again, this is in qualitative agreement with the results in Fig.3.

RESULTS IN QUADRANT I

Let us now come to the measurements taken in the first quadrant. The sample we have investigated is a mixture of water and ethanol with a 10% concentration by weight of ethanol. The system is characterized by a small negative value of S, and it is heated from below.

The system now is overstable, due to the stabilizing effect of the concentration gradient (ethanol, the lighter component, migrates to the top plate). The threshold is slightly higher than for the RBI (see Fig.1), and one expects to observe oscillations for all the three modes, the temperature, the concentration and the velocity mode. For free free, previous boundaries the stability analysis gives analytic expressions for the frequency of the critical mode (4,5).

We will not report here the exact solution and will give an approximate expression

$$\omega_{oscill} \simeq \frac{6 \; \pi^2 \; \chi \; \left[-S/(1+S)\right]^{1/2}}{4 \; a^2} \tag{16}$$

For the more realistic case, rigid rigid impervious boundaries computations yield slightly larger estimates (4,5) while the critical wavevector is

$$k_c = (2\pi/\Lambda) \sim (2\pi/2.02 \; a) \tag{17}$$

In view of the poor quality of the data available for S for the ethanol water system, we will use Eq.16 for a rough estimate and comparison with experimental data.

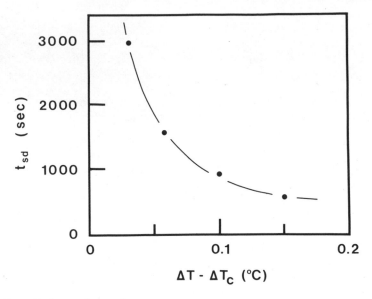

Fig. 5.   Relaxation time constants for the approach to steady
          state (Polystyrene sample).

     The preliminary observations in the quadrant I were performed
in order to ascertain the oscillatory behaviour of the variables
in the pretransitional region by applying a stimulus to the system
and then watch the decay to the steady state. We decided to
apply the stimulus in the form of heat pulse delivered to a small
fluid volume in the center of the cell. This was achieved by
slightly modifying one of our cells. A cylindrical rod of 2 mm
in diameter of a black insulating material was inserted in a hole
practiced in the center of the lower plate. The insert was then
mechanied flush to the surface. Heat pulses were applied by
bringing an argon laser beam onto the absorping spot for a few
seconds. By operating in this way, the system is allowed to
select spontaneously both the wavevector and the frequency of the
oscillations which, at least close to threshold, should be in
agreement with Eqs.16, 17. The fluid sample is 5 mm high and
20 mm in diameter, and the heating laser beam is brought onto
the absorbing spot at a skew angle across the cylindrical glass
window. The observation of the oscillations is made with the
laser beam deflection technique. As in the cases discussed before,
a low power He-Ne laser beam is mildly focused in the midplane of
the fluid sample. In order to understand how this technique can
be applied in this case we must discuss a few points. Let us
recall that although concentration and temperature modes introduce

spatial variations of the refractive index, they do not produce per
se any refractive index gradient once one takes the average over
the length of the sample. Gradients, and therefore beam deflections
arise because we also have a velocity mode, which coupled to the
other two produces a mass flow (like in the cases discussed
before) or heat flow (like in the R.B. instability), or both,like
in the case under study. The flows are the ones which are
responsible for the reduction of the average concentration and
temperature gradients, and ultimately of the beam deflection.

As we said all the three variables, velocity temperature and
concentration must be considered in this case. If one takes
for the perturbation in the vertical component of the velocity
(assuming for simplicity a rectangular geometry) the expression

$$v_z = f(z) \cos k x \cos \omega t \tag{18}$$

then the associated temperature and concentration perturbations
take the form (10)

$$\delta T = g(z) \cos k x (A_1 \cos \omega t + B_1 \sin \omega t)$$

and $\tag{19}$

$$\delta c = h(z) \cos k x (A_2 \cos \omega t + B_2 \sin \omega t)$$

where A_1, B_1, A_2, B_2 are ω dependent quantities.

Eqs. 18, 19 indicate that the relative phases between the
modes depend on the frequency. Mass and heat flows due to
convection are proportional to $v_z \, \delta c$ and $v_z \, \delta T$, and therefore
they will oscillate at a frequency 2ω, since terms of the type
$\sin \omega t \cos \omega t$ or $\cos^2 \omega$ will appear when one multiplies Eq.18 by
Eq.19. Consequently gradient losses and therefore beam deflections
will also oscillate at frequency 2ω.

Measurements were taken by reaching a desired temperature
difference in small steps and then releasing the pulse and
monitoring the time evolution of the beam deflection. The
approach in small steps is necessary in order to let the stabilizing
concentration gradient build up. The small step approach is
essential especially close to the threshold above the RBI
threshold. Should such a temperature difference be applied
suddenly, the system could then become prematurely unstable
because the stabilizing effect of the solute concentration would
not have a chance to develope.

Measurements have been taken at three temperature differences
close to the threshold (see Fig.6). The energy indicated in
the figures is the laser beam energy. Only a few percent of this
energy is actually absorbed in the black spot, due to the grazing
angle of incidence of the incoming beam. Oscillations appear only

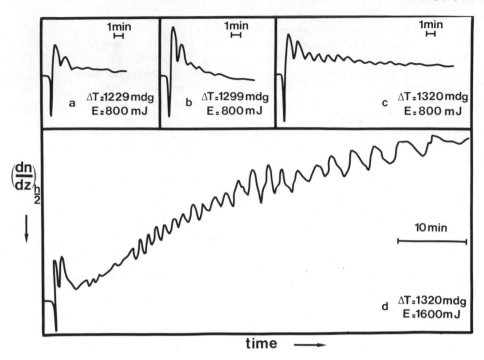

Fig. 6.   Stimulated oscillatory behaviour for the water-ethanol
          system.

fairly close to threshold. Qualitatively one notices that the
frequency of the oscillations does not change appreciably, and
the period is roughly 60 sec. The period of the modes should then
be about 120 sec. As threshold is approached, the oscillations
become more persistent, as expected since exactly at threshold
they should remain undamped. For the last case we have also
applied twice the power used before. This is enough to make
the system flip over to the unstable state (see Fig. 6d). The
gradient now, instead of falling back to its original stable
value evolves via oscillations toward a fully developed convective
instability. Notice also the slow change in frequency of the
oscillation. The interpretation of Fig.6d is that the heat pulse
is now large enough to make oscillations so ample to actually
overturn locally the solute distribution. As a consequence of
these large amplitude oscillations, the stabilizing solute
gradient is slowly destroyed and stationary convection sets in.

Furthermore, the qualitative change in frequency may perhaps
be interpreted as a consequence of the gradual weakening of the
stabilizing effect of the concentration gradient.

In concluding let us try to extract at least one piece of quantitative information from these data and compare it with the theory and other published data. The frequency of the mode oscillations we obtain is $\omega \simeq 5.2 \times 10^{-2}$ sec^{-1} and therefore $\omega a^2 \simeq 13 \times 10^{-3}$ cm^2/sec. This has to be compared with the value for ωa^2 obtained in Ref.4 for the system methanol water at the same concentration $\omega a^2 \simeq \&.4 \times 10^{-3}$ and the value 7.3×10^{-3} obtained in Ref. 5 for the system water ethanol at the same concentration. So, although there is an order of magnitude agreement, our results differ almost a factor of two from those reported in the literature.

A comparison of experimental results in Ref.4 and Ref.5 with the prediction based on Eq.16 and using available data for S indicates that the agreement is fair, although for the methanol water case Eq.16 predicts $\omega a^2 = 9.5 \times 10^{-3}$ in closer agreement with our data.

As a tentative explanation for discrepancy of our results and those in the literature one may consider that due to the highly localized application of the heat pulse, the wavevector excited is somewhat higher than that predicted by the theory. Also, by carefully checking the thermal properties of the low aspect ratio cell used for these measurements we discovered that the temperature gradient prior convection does not remain constant as a function of height and variations between center and edges amount to several percent. As improved cell is now under construction and new data are expected in the near future.

This work was supported by CNR/CISE contract no.7800901.02.

REFERENCES

1. S. R. de Groot and P. Mazur, "Non Equilibrium Thermodynamics", North Holland, Amsterdam (1956).
2. M. Giglio and A. Vendramini, Phys. Rev. Lett., 39,1014 (1977).
3. M. Giglio and A. Vendramini, to be submitted for publication.
4. D. T. Hurle and E. Jakeman, J. Fluid Mech., 47:667 (1971).
5. M. G. Velarde and R. S. Schechter, Phys. Fluids, 15:1707 (1972).
6. J. K. Platten and G. Chavepeyer, J. Fluid Mech., 60:305 (1973).
7. M. Giglio and A. Vendramini, Phys. Rev. Lett., 38:26 (1977).
8. J. A. Bierlein, J. Chem. Phys., 23:10 (1955).
9. V. Degiorgio, Phys. Rev. Lett., 41:1293 (1978).
10. H. N. W. Lekkerkerker, Physica, 93A:307 (1978).

THERMODYNAMIC ANALYSIS OF INVERTED BIFURCATION

H. N. W. Lekkerkerker

Faculteit van de Wetenschappen
Vrije Universiteit Brussel
Belgium

ABSTRACT

We present a thermodynamic analysis of inverted bifurcation
in binary mixtures heated from below. From this analysis it
follows that an inverted bifurcation is caused by the competition
between a stabilizing effect with a long relaxation time and a
destabilizing effect with a short relaxation time. These
conditions are precisely the same as those that give rise to
overstability. This might explain why overstability and inverted
bifurcation occur in the same systems.

INTRODUCTION

The onset of convection in a layer of pure liquid heated from
below, the so-called Bénard-Rayleigh instability, has been
extensively investigated for a long time (for reviews see Refs.
1 and 2). In recent years there has been considerable interest
in thermal convection effects in horizontal layers of binary
mixtures (3) and nematic liquid crystals (4). The Bénard-Rayleigh
instability in these systems exhibits features that are dramatically
different from those observed in pure isotropic liquids. The
most spectacular effects have been observed in binary mixtures
where the Soret effect causes the more dense component to move to
the warm boundary and in homeotropic nematics with a positive heat
conduction anisotropy. These systems become unstable to
stationary convection when heated from above even though the
overall density gradient is not adverse (5,6). Furthermore when
heated from below these systems become unstable to oscillatory
convection (overstability) and finite amplitude convection
(7,9) (inverted bifurcation).

231

Previously we pointed out that overstability is due to the competition between a stabilizing effect with a long relaxation time and a destabilizing (10,11) effect with a short relaxation time. The fact that overstability and inverted bifurcation occur together in such disparate systems as binary mixtures and nematic liquid crystals suggests that inverted bifurcation is also due to the difference in time scales between the stabilizing and destabilizing effect. In this paper we show that this is indeed the case for a binary mixture.

Overstability and inverted bifurcation in binary mixtures have been studied numerically by Platten and Chavepeyer (12,13) and quite recently also by Velarde and Antoranz (14,15). The aim of the present work is not so much quantitative but rather to elucidate the basic physical mechanism that gives rise to an inverted bifurcation.

CLASSIFICATION OF INSTABILITIES

A convenient starting point for the classification of instabilities is the Landau expansion (16) of the rate of change of the kinetic energy of a convective disturbance.

$$E_{kin} = a(R-R_c)v^2 + Bv^4 + Cv^6 + \ldots \qquad (a > 0) \qquad (1)$$

Here v is the amplitude of the convective disturbance, R is a parameter characterizing the non-equilibrium constraints on the system (Reynolds number, Rayleigh number, Taylor number ...) and R_c is the value of this parameter for which the system becomes unstable (critical value). In the stationary state there is a balance between the rate of injection of energy into the convective disturbance and the rate of viscous dissipation of kinetic energy associated with the convective disturbance

$$\dot{E}_{kin} = 0 \qquad (2)$$

In case the coefficient B is negative and large the higher order terms in the expansion (1) are irrelevant and one obtains from (2)

$$v = 0 \qquad\qquad \text{for} \qquad R < R_c$$

$$V = \text{const.} (R-R_c)^{1/2} \qquad \text{for} \qquad R > R_c$$

This case is commonly referred to as a direct or normal bifurcation and is analogous to a second order (continuous) phase transition. The behaviour of v as a function of R is schematically represented in fig. 1. The above theoretical prediction for the behaviour of

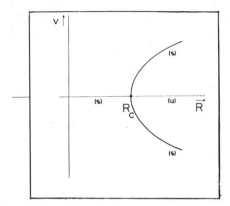

Fig. 1.   Schematic representation of the variation of the amplitude
          of the velocity for a direct bifurcation
          (s) stable   (u) unstable

the amplitude of the convective disturbance, originally due to
Landau, has been experimentally verified for the Bénard–Rayleigh
instability (17) and the Taylor instability (18) using light
scattering techniques.

     A different situation arises when B is positive. Now the
coefficient C must be negative to ensure that (2) can be
satisfied. In this case (2) has already solutions $v \neq 0$ for
$R \geqslant R_o$ where

$$R_o = R_c - \frac{B}{4a|C|}$$

(subcritical finite amplitude instability) and the system
exhibits a hysteresis loop between R_o and R_c (see fig.2). This
case is known as inverted bifurcation and is analogous to a
first order (discontinuous) transition.

     The cross over between direct and inverted bifurcation takes
place for B = 0. In this case

$$v = 0 \qquad\qquad \text{for} \qquad R < R_c$$

$$v = \text{const.}(R-R_c)^{1/4} \qquad \text{for} \qquad R \geqslant R_c$$

This situation is analogous to a tricritical point in equilibrium
phase transitions (19).

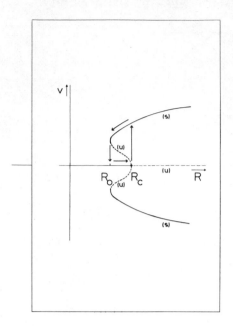

Fig. 2.   Schematic representation of the variation of the amplitude
           of the velocity and hysteresis loop for an inverted
           bifurcation
           (s)   stable   (u)   unstable

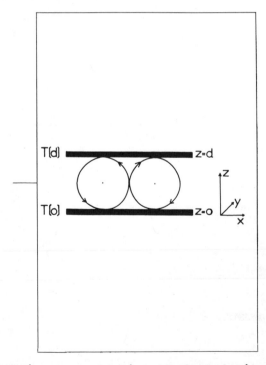

Fig. 3.   Schematic representation of the velocity perturbation.

and d is the thickness of the fluid layer. As a preliminary to
the calculation of E_{kin} for a binary mixture we first consider
a pure fluid. The rate of viscous dissipation of kinetic energy
per unit volume is given by

$$\dot{E}^v_{kin} = -\frac{1}{V}\frac{1}{2}\eta \int (\frac{\partial v_i}{\partial x_j} + \frac{\partial v_j}{\partial x_i})^2 \, dV$$

$$= -2\eta q^2 v^2 \tag{4}$$

where

$$q^2 = q_x^2 + q_z^2$$

and η is the shear viscosity. The rate of production of kinetic
energy by the buoyancy force per unit volume is given by

$$\dot{E}^g_{kin} = -\frac{1}{V} \int g \, \delta\rho \, v_z \, dV \tag{5}$$

where g is the gravitation constant and $\delta\rho$ is the perturbation
in the mean steady-state density distribution. For a pure
fluid we can write

$$\delta\rho = -\rho \, \alpha \, \delta T \tag{6}$$

where δT is the perturbation in the mean steady-state temperature
distribution and α is the thermal expansivity. The perturbation
δT is determined by

$$v_z \frac{d<T>}{dz} = \chi \nabla^2 \delta T \tag{7}$$

where χ is the thermal diffusivity and $<T>$ is the mean steady-state
temperature distribution ($<...>$ denotes an average over the
horizontal plane). Following the work of Chandrasekhar (22) one
obtains to order v^2 for the mean steady-state temperature
distribution

INVERTED BIFURCATION IN BINARY MIXTURES

     The fundamental physical process that lies at the origin of
the Bénard-Rayleigh instability is the conversion of energy released
by the buoyancy force into kinetic energy of the convective motion
(20). Stationary convection sets in when the rate of injection
of energy by the buoyancy force acting on the fluid (E^g_{kin})

begins to balance the rate of viscous dissipation of energy
$(\dot{E}^V_{kin})$

$$\dot{E}_{kin} = \dot{E}^V_{kin} + \dot{E}^g_{kin} = 0$$

when the convective disturbance has a finite amplitude, convective
transport causes a modification of the mean (horizontally-
averaged) temperature and concentration distribution. This in turn
modifies the rate of liberation of thermodynamically available
energy by the buoyance force acting on the fluid (21,22).

For the sake of simplicity we consider a convective
disturbance in the form of a circular roll pattern (see Fig.3).

$$v_x = - 2v \sin q_x x \cos q_z z$$

$$v_y = 0 \tag{3}$$

$$v_z = 2v \cos q_x x \sin q_z z$$

Here

$$q_x = q_z = \frac{\pi}{d}$$

$$\frac{d<T>}{dz} = - \beta \{1 + \frac{v^2}{\chi^2 q^2} \cos 2 q_z z\} \tag{8}$$

where

$$\beta = \{T(z=0) - T(z=d) \} /d$$

The second term on the r.h.s. of (8) represents the modification
of the temperature gradient due to convective heat transport.
Substituting the temperature distribution (8) in (7) one obtains

$$\delta T = \frac{\beta}{\chi q^2} 2v \{\cos q_x x \sin q_z z - \frac{v^2}{\chi^2 q^2} (\frac{1}{5} \cos q_x x \sin q_z z +$$

$$\frac{2}{5} \cos q_x x \sin^3 q_z z)\} \tag{9}$$

Using (9) in (6) one obtains for the rate of production of
kinetic energy by the buoyancy force

$$\dot{E}^g_{kin} = \frac{g \rho \alpha \beta v^2}{\chi q} \{ 1 - \frac{v^2}{2\chi^2 q^2} \} \tag{10}$$

where δc is the perturbation in the mean steady-state concentration distribution and α' is the solutal expansivity. The perturbation δc is determined by

$$v_z \frac{d <c>}{dz} = D \nabla^2 \delta c \qquad (13)$$

where D is the mass diffusion coefficient and $<c>$ is the mean steady-state temperature distribution. Using the same line of reasoning as used in the calculation of the mean temperature distribution one obtains

$$\frac{d<c>}{dz} = \frac{k_T \beta}{T} \{1 - \frac{2v^2}{D^2 q^2} \sin^2 q_z z\} \qquad (14)$$

where k_T is the thermal diffusion ratio. The first term on the r.h.s. of (14) represents the Soret driven concentration gradient and the second term represents the modification of this concentration gradient due to convective mass transport. Substituting Combining (4) and (10) one obtains for the rate of change of the kinetic energy of the convective disturbance (3)

$$\dot{E}_{kin} = a(R-R_c)v^2 + Bv^4 \qquad (11)$$

where

$$a = \frac{\eta}{d^4 q^2}$$

$$R = \frac{g \alpha \beta d^4}{\chi \nu} \qquad \text{(Rayleigh number)}$$

$$R_c = 2q^4 d^4 \qquad \text{(critical Rayleigh number)}$$

$$B = - \frac{g \rho \alpha \beta}{2 \chi^3 q^4}$$

Here ν is kinematic viscosity. We see that B is negative and thus at $R = R_c$ the system will undergo a continuous transition to the convective state (direct bifurcation).

In the case of a binary mixture the perturbation in the mean steady-state density distribution can be written as

$$\delta \rho = - \rho \alpha \delta T + \rho \alpha' \delta c \qquad (12)$$

the concentration distribution (14) in (13) one obtains

$$\delta c = - \frac{\frac{k_T}{T} \beta}{Dq^2} \; 2v \; \{\cos q_x x \sin q_z z - \frac{v^2}{D^2 q^2} (\frac{6}{5} \cos q_x x \sin q_z z$$

$$+ \frac{2}{5} \cos q_x x \sin^3 q_z z)\} \tag{15}$$

Using (9) and (15) in (12) one now obtains for the rate of production of kinetic energy by the buoyancy force

$$\dot{E}^g_{kin} = \frac{g\rho\alpha\beta}{q^2} (\frac{1}{\chi} + \frac{S}{D}) v^2 - \frac{g\rho\alpha\beta}{2q^4} (\frac{1}{\chi^3} + \frac{3S}{D^3}) \; v^4 \tag{16}$$

The dimensionless parameter

$$S = \frac{k_T}{T} \frac{\alpha'}{\alpha}$$

is negative in case the heavy component moves to the warm boundary. Combining (4) and (16) one obtains for the rate of change of the kinetic energy of the convective disturbance (3) in a binary mixture

$$\dot{E}_{kin} =\cdot a(R-R_c)v^2 + Bv^4$$

where

$$a = \frac{\eta}{d^4 q^2} \; (1 + \frac{S\chi}{D})$$

$$R = \frac{g \; \alpha\beta \; d^4}{\chi\nu}$$

$$R_c = 2q^4 d^4 \frac{1}{1 + \frac{S\chi}{D}}$$

$$B = - \frac{g\rho\alpha\beta}{2q^4} (\frac{1}{\chi^3} + \frac{3S}{D^3})$$

It is clear that for binary mixture heated from below where the more dense component moves to the warm boundary (S < 0) the Soret effect to lowest order in v has a stabilizing influence i.e. it leads to an increase of R_c. Let us now consider the coefficient B, the sign of which determines whether we are dealing with a direct bifurcation (B < 0). The coefficient B contains two distinct contributions of opposite sign. The negative contribution

$$- \frac{g\rho\alpha\beta}{2\chi^3 q^4}$$

is due to the fact that convection changes the mean temperature distribution in such a way that the rate of transfer of energy from the gravitational field to the convective disturbance is lowered. The positive contribution

$$\frac{-3g\rho\alpha\beta}{2D^3 q^4}$$

is due to the fact that convection stirs up the Soret driven concentration gradient there by lowering its stabilizing effect. Since for liquid mixtures $D \ll \chi$ the effect of convection on the stabilizing concentration gradient is much larger than on the destabilizing temperature gradient. The result is that for values of S smaller than $\frac{1}{3} \frac{D^3}{\chi^3}$ the positive term in B dominates and thus for these values of S the system will exhibit an inverted bifurcation.

CONCLUSION

The inverted bifurcation in a binary mixture heated from below where the more dense component moves to the warm boundary is due to the competition of a stabilizing effect with a long relaxation time and a destabilizing effect with a short relaxation time. Precisely the same conditions give rise to overstability. Actually oscillatory convection and finite amplitude convection lead in different ways to the same result i.e. both effectively eliminate the slow stabilizing effect while retaining the fast destabilizing effect.

It would be of interest to study the cross over between direct and inverted bifurcation experimentally. In binary liquids D/χ is typically 10^{-2} and thus the cross over takes place for $S \simeq - 10^{-6}$ which is experimentally hard to realize. A system that may offer better chances to perform the appropriate experiments is a homeotropic nematic with positive heat conduction anisotropy

heated from below. Here it is possible to change the ratio of the
relevant relaxation times with a stabilizing magnetic field (9)
which could induce the cross over between direct and inverted
bifurcation.

ACKNOWLEDGEMENT

This work was initiated by stimulating discussions with
E. Guyon.

REFERENCES

1. S. Chandrasekhar, Hydrodynamic and Hydromagnetic Stability
    Clarendon Press, Oxford (1961).
2. C. Normand, Y. Pomeau and M. G. Velarde, Rev. Mod. Phys.,
    49:581 (1977).
3. R. S. Schechter, M. G. Velarde and J. K. Platten, Adv. Chem.
    Phys., 26:265 (1974).
4. E. Dubois-Violette, G. Durand, E. Guyon, P. Manneville and
    P. Pieranski, Solid State Physics, Supplement 14: 147 (1978).
5. R. S. Schechter, I. Prigogine and J. R. Hamm, Phys. Fluids,
    15:379 (1972).
6. E. Dubois-Violette, E. Pieranski and E. Guyon, Phys. Rev.
    Lett., 30:736 (1973).
7. D. T. J. Hurle and E. Jakeman, J. Fluid Mech., 47:667
    (1971).
8. J. K. Platten and G. Chavepeyer, J. Fluid Mech., 60:305
    (1973).
9. E. Guyon, P. Pieranski and J. Salan, J. Fluid. Mech., 93:65
    (1979).
10. H. N. W. Lekkerkerker, J. Phys. Lett., (Paris) 38:L-277
    (1977).
11. H. N. W. Lekkerkerker, Physica, 93A:307 (1978).
12. J. K. Platten and G. Chavepeyer, Int. J. Heat Mass Transfer,
    18:1971(1975).
13. J. K. Platten and G. Chavepeyer, Int. J. Heat Mass Transfer,
    20:113 (1977).
14. M. G. Velarde in Dynamical Critical Phenomena and Related
    Topics, C. Enz. ed., (Lecture notes in Physics 104, Springer
    Verlag) (1979) p. 309.
15. M. G. Velarde and J. C. Antoranz, Phys. Lett., 72A:123
    (1979).
16. L. Landau and E. Lifchitz, Mécanique des Fluides (Editions
    Mir, Moscou) (1971) Sect. 26.
17. P. Berge in Fluctuations, Instabilities and Phase
    Transitions, T. Riste ed., Plenum, New York (1975) 323.

18. J. P. Gollub and M. H. Freilich, Phys. Rev. Lett., 33:1465
     (1974).
19. R. B. Griffith, Phys. Rev. Lett., 24:715 (1970).
20. S. Chandrasekhar, Max Planck Festschrift 1958 (Veb.
     Deutscher Verlag der Wissenschaften, Berlin, 1958) 103.
21. J. T. Stuart, J. Fluid Mech., 4:1 (1958).
22. S. Chandrasekhar, Ref. 1 Appendix I.

FORCED RAYLEIGH LIGHT SCATTERING IN FLUIDS

F. Rondelez

Collège de France, Physique de la Matière
Condensée
11 Place M. Berthelot 75231 Paris Cedex 05
France

ABSTRACT

Various applications of the forced Rayleigh light scattering technique in fluids are reviewed: thermal conductivity, translational and rotational mass diffusion, hydrodynamic velocity gradients, flash photolysis. Comparison is made with the quasi-elastic light scattering technique when applicable. Particular emphasis is given on the advantages of selectively inducing a spatially-periodic modulation of the relevant quantity to be measured. However, it will be shown on a particular example that these two light scattering techniques are certainly complementary. Several new developments are also proposed.

1. INTRODUCTION

Dynamical studies in fluids have been revolutionarized in the mid-sixties by the advent of the quasi-elastic Rayleigh light scattering (QELS). Excellent reviews (1,2,3) describe the technique, which relies on the analysis of the broadening of the central Rayleigh line by thermodynamic statistical fluctuations. In its hey-days, QELS has been applied not only to simple fluids (4) but also to binary mixtures, normal or critical (5), to solutions of macromolecules (6) and even to liquid crystals (7). In spite of this success, the technique was suffering from several drawbacks. First, the amplitude of the spontaneous, statistical, fluctuations is small and this severely limits the sensitivity of the system. Second, all the thermodynamic fluctuations are intertwined and several parameters such as density, temperature, concentration can be excited together, obscuring each other in some cases (a well-

known example is the entropy line in critical binary mixtures which
is completely hidden by the Rayleigh component arising from the
concentration fluctuations). Third, although QELS allows to study
spectral lines far too narrow to be measured by convential
spectroscopic techniques, its frequency resolution is still not
sufficient to study slow relaxation processes with decay times
much above one second.

In an attempt to overcome these difficulties, several groups
working independently (8,9,10) started in the early seventies to
investigate the possibility of photo-inducing strong, coherent,
spatially periodic temperature fluctuations in weakly absorbing
media. The amplitude of these fluctuations was much larger than
those of the statistical fluctuations but also low enough to always
stay close to thermal equilibrium. The name of forced Rayleigh
light scattering (FRS) was coined to this method (10). Thus the
analogies with QELS was emphasized on one hand but the distinction
with stimulated Rayleigh scattering (SRS), which is based on nonlinear effects (11) absent in FRS, is clearly made on the other.
In many respects, FRS is simpler than QELS and this has prompted
numerous experiments. A first review of the subject has been
recently given by Pohl (12). Here we shall mainly concentrate
on the application of FRS to fluids. In section 2, we first
describe briefly the principle of the method and the experimental
set-up. Then we show its application to heat transport, mass
transport (both translational and rotational), flash photolysis,
hydrodynamic measurements in sections 3, 4, 5, 6, respectively.
In section 7, we will conclude and give an outlook of possible
future developments.

2.   PRINCIPLE OF FRS

The basis of FRS is to replace the weak statistical thermal
fluctuations of QELS by strong, coherent fluctuations induced
externally. The thermodynamic parameter of interest - temperature,
concentration, molecular orientation, etc... - is spatially
modulated with a modulation depth large compared to thermal
fluctuations but weak enough to stay close to equilibrium. In
the case of a temperature perturbation, for instance, a change of
10^{-3} ^{o}K will not perturb the system significantly but still is six
orders of magnitude greater than the $<\Delta T^2>^{1/2} \simeq 10^{-9}$ ^{o}K induced
by the statistical fluctuations (13). Since this external
perturbation is spatially modulated, it can be easily observed
through the concomitant refractive index change as an optical grating
with well-defined diffraction properties. The light of an incident
monitoring laser will be diffracted into angles directly related to
the wavevector $\vec{q_o}$ of the spatial modulation. The light intensity
will be proportional to the square of the refractive index change,
at least to first order. Both variations in the real and imaginary
parts of the refractive index will contribute. Such optical gratings

are called phase and amplitude gratings respectively and have been extensively studied in holographic techniques (14). Following the perturbation impulse, the diffracted intensity rises very quickly and then decays exponentially back to its original zero value if the processes involved are purely diffusive or relaxational. From the spatial dependence of the characteristic decay time, it is possible to separate between a purely intra-molecular relaxation or a transport phenomena.

The spatially-varying, amplitude controlled, external modulation can be achieved by various methods. The most popular so far has been to use the sinusoîdal interference pattern generated when two beams issued from the same laser are crossed under an angle θ. The relation between θ, λ_o the excitation laser wavelength, and q_o the wavevector of the interference pattern, is :

$$\left| q_o \right| = \frac{4\pi}{\lambda_o} \sin \frac{\theta}{2}$$

The pattern is envolope modulated by the Gaussian profile of the laser beam, giving it finite extent. The sinusoîdal excitation pattern interacts with the sample through various means. If the sample is absorbing at the incident wavelength, it will induce a periodic temperature rise. If the sample is photochromic, it may also create directly photoexcited states. Although, the possibility of a contact-free, optical perturbation is very attractive it should be borne in mind that other spatially-alternating sources can be used such as electric fields applied between interdigital electrodes, standing acoustic waves, etc...

The standard experimental set-up is shown on figure 1. It involves two lasers operated at generally different wavelengths λ_o and λ_1 and used as pump and probe respectively. The intense argon or krypton ion laser beam is mechanically chopped to give light pulses of 40 µs duration with a repetition time variable between 20 ms and several minutes. This beam is then split by a beam-splitter mirror arrangement into two beams of equal intensities, I_1 and I_2. There is a small angle between the two beams so they can interfere under an angle θ into the sample. In this two-beam interferometer, the optical interference pattern is an array of successive bright and dark fringes, the intensity of which is sinusoîdally modulated in space. A real image of the interference pattern can be made by a microscope objective onto a distant screen to determine the fringe spacing $1 = 2\pi/q_o$. The orientation of the linear fringe pattern relative to the sample can be changed continuously by rotating the Dove prism.

We have already mentioned that the sample can interact with the pump beams by a variety of means. In each case, this will induce a spatially-periodic change in the refractive index of the

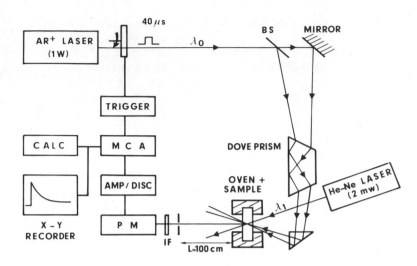

Fig. 1. Experimental set-up, taken from reference (25).

medium. The resulting phase or amplitude grating is detected
by the second laser whose output is diffracted by the grating. The
intensity of this probe beam is weak enough not to perturb the
sample under examination. Its angle of incidence is such to
satisfy the Bragg conditions. Indeed it should be remembered that
although the interference pattern is periodic in one direction, it
is approximately uniform along the two others (we assume here
that the sample is only weakly absorbing). The diffraction grating
is thus tri-dimensional. The angular conditions become very
critical when the sample thickness d is large. If d is ≃ 5mm,
Bragg diffraction angles have to be adjusted within 2 arc minutes.
The diffracted pattern is filtered through a small pin-hole and
only the first order diffraction dot is allowed to reach the
photomultiplier tube through an optical filter centered at the
probing wavelength λ_1. The electrical output is amplified and fed
into a multichannel analyzer used in the multiscaler mode.
The time dependence of the diffracted intensity is recorded after
each exciting pulse and averaged over many periods to improve the
accuracy. The resulting data are then processed with a small
computer in order to extract the relaxation time (or times) of
the diffracted light intensity.

     The time analysis of the diffracted intensity is only correct
if. coherence properties of the scattered electric fields $E_s(t)$
are taken into account. All the various $E_s(t)$ emitted by the
illuminated parts of the sample volume add up only if they are
collected on the phototube over less than one coherence area (1).
In that case, the output voltage $V_a(t)$ is proportional to
$\left| \sum_n E_{s,n}(t) \right|^2$. In the summation, we have to include the electric

fields B due to the unavoidable static defects, which give a non-zero background in the direction of observation. Depending on their amplitude relative to the signal term, the detection scheme is called heterodyne or homodyne. If the two are comparable, it is often convenient to add an external local oscillator so to avoid to be in a mixed regime always difficult to analyze. $V_a(t)$ writes:

$$V_a(t) \quad \propto \quad (E_s(t) + B)^2$$

If $B \gg E_s(t)$, one has $V_a(t) \quad \propto \quad BE_s(t) + B^2$

The time dependent part of $V_a(t)$ is directly related to the time dependence of the signal term $E_s(t)$.

If $B \ll E_s(t)$, $V_a(t) \quad \propto \quad \left| E_s(t) \right|^2$

$V_a(t)$ is related to the square of the signal term. An ingenious detection scheme to eliminate spurious dynamic effects due to the finite size of the grating pattern has been developed by Chan and Pershan (15).

3.   THERMAL TRANSPORT

FRS has been extensively applied to thermal diffusivity measurements in fluids. In many respects, it is more satisfying than the previously existing methods. The absence of physically perturbing probes is a decisive advantage over the classical hot wire method. The possibilities of using small sample volumes, of controlling accurately the applied temperature gradients and of overcoming other fluctuating sources such as concentration fluctuations in binary mixtures, are also of importance over the heat flux measurement method, the thermal lens effect (16) and the QELS technique (1) respectively. We will now describe in details the application of FRS to thermal transport since many of the features will be applicable to other sections of this paper.

In all thermal experiments, the two-beam interference pattern is converted into a temperature grating by partial absorption of the incident light. Transparent samples are made absorbing by admixture of soluble dyes (9). The temperature grating is periodic in space, with the same chacteristics as the interference pattern.

$$T(x,t) = T_o + \delta T(t) \cos q_o x$$

The use of a weak, sinusoîdal, modulation of temperature T is convenient because it is self preserving : if the profile $T(x)$ is sinusoîdal at time $t = 0$, it will remain sinusoîdal at later times and with the same wavevector q_o. Its amplitude will decrease with time as given by the diffusion equation for heat

$$\rho C_P \frac{\partial T}{\partial t} - K \frac{\partial^2 T}{\partial x^2} = 0 \tag{1}$$

K is the heat conductivity; ρ is the density; C_p is the specific heat at constant pressure. Here we have considered a one-dimensional diffusion, which is correct if the extension of the grating in the two other directions is large compared to the fringe spacing $1 = 2\pi/q_o$.

The solution to the diffusion equation is

$$T(x,t) = T_o + \delta T(o) \exp{-\frac{t}{\tau}} \cos q_o x \tag{2}$$

The modulation depth $\delta T(t)$ decreases exponentially with a time constant τ

$$\tau = \frac{\rho C_P}{Kq^2} \tag{3}$$

The maximum temperature modulation $\delta T(o)$ is obtained at the end of the exciting pulse of duration t_p

$$\delta T(o) = \frac{\alpha(I_1 I_2)^{1/2} t_p}{\rho C_P} \tag{4}$$

α is the absorption coefficient of the sample at the frequency of the pump beams of intensities I_1 and I_2 respectively (generally $I_1 = I_2$).

The spatially periodic temperature distribution generates a periodic modulation of the refractive index

$$n(x) = n_o + \delta n \cos q_o x \tag{5}$$

which acts as a phase grating. The phase variation of the grating is:

$$\delta\emptyset = \frac{2\pi d}{\lambda_1} \left(\frac{dn}{dT}\right)_P \delta T \tag{6}$$

The light intensity diffracted by this grating is, in the limit of small phase variation $\delta\emptyset \ll 1(14)$,

$$\frac{I_{diff.}}{I_{inc.}} = \frac{1}{4} T_{\lambda_1} \phi^2 \tag{7}$$

$I_{inc.}$ is the incident intensity of the probe beam. T_{λ_1} is the sample transmission at the monitoring wavelength λ_1. No local oscillator have been considered here. The light intensity will decrease exponentially with time as

$$I_{diff.}(t) \propto \exp \frac{-2t}{\tau} \tag{8}$$

The characteristic relaxation time of the diffracted light intensity is then

$$\tau_1 = \frac{\tau}{2} = \frac{\rho C_p}{K q_o^2} = \frac{1}{D q_o^2} \tag{9}$$

D is the thermal diffusivity, $D = K/\rho C_p$. It is easily measured experimentally from the slope of the curve τ_1 versus q_o^2, see figure 2.

The thermal diffusivity is a combination of thermal conductivity, specific volume and heat capacity. It is possible to get the thermal conductivity alone by combining the amplitude and the relaxation time measurements, see eq.4 and 9.

This is no longer a problem in anisotropic materials such as liquid crystals in which the most important quantity is the ratio between the thermal conductivities parallel and perpendicular to the director axis. In that case, $K_{//}/K_\perp = D_{//}/D_\perp$. FRS has allowed systematic measurements on various liquid crystals in both the nematic abd smectic phases since it requests only small sample volumes, easy to align macroscopically (17,18,19). A nice outcome was the understanding of the origin of the thermal anisotropy in these phases consisting of strongly elongated molecules. Contrary to previous beliefs, the main factor is the geometrical shape of the molecules and not the positional ordering of their centers of mass(18). This explains the perfect continuity of the thermal diffusion data between the smectic and nematic phases of a same material (17). This also explains why the thermal anisotropy increases when the central rigid core of the molecules is extended, see figure 2. In smectic phases, the thermal diffusivity has also been measured at all angles relative to the smectic layers. The corresponding polar plot has a striking peanut-shape displayed on figure 3.

Recently Pohl (20) has applied FRS to thermal diffusivity measurements in critical binary mixtures. He was able to extract from his data the critical exponent for D. $D \propto (T - T_c)^{-x}$ with

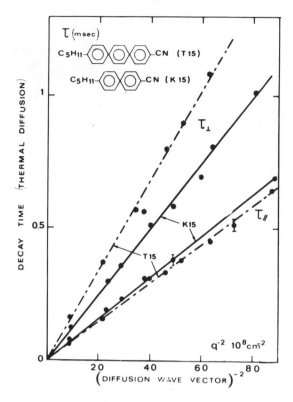

Fig. 2.   Taken from reference (18). Decay time fo the thermal
          grating vs the inverse square of the grating wavevector
          for two nematic liquid crystals. (K15)4'-n-pentyl-4-
          cyanobiphenyl, T(15)4'-n-4-cyanoterphenyl. In each
          case, two relaxation times $\tau_{//}$ and $\tau_\perp$ are given for the
          two principal molecular orientations relative to the fringe
          pattern.

x = 0.58.   It is certainly an interesting domain of investigation
since QELS is pleagued in that case by the divergent concentration
fluctuations (critical opalescence). Other examples of thermal
transport studies are given by Chan and Pershan (21) for lipidwater
smectic phases and by Lallemand et al (22) for a relaxing fluid as
glycerol.

     So far we have only considered diffusive thermal transport.
A remarkable propagating thermal mode has been recently predicted
and observed by Boon, Allain and Lallemand (23) in fluids under
strong thermal constraints. They have studied hydrodynamically
stable fluid layers submitted to upward-directed temperature
gradients. Only a finite amount of heat can be transported by
thermal diffusion and above a critical value of the temperature

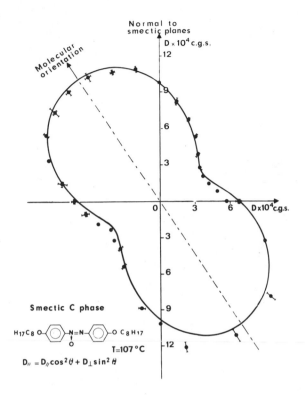

Fig. 3.   Taken from reference (18). Angular dependence of the
          thermal diffusivity with respect to the average molecular
          orientation and to the smectic layers for a liquid
          crystalline material. Smectic C phase of 4-4' di-n-
          octyloxy azoxy benezene.

gradient the heat must propogate as a thermal wave with a well-
defined velocity and damping time constant. FRS has been used to
excite a standing horizontal temperature pattern in thin fluid
layers of acetone contained between two temperature-controlled
glass plates. The diffracted light intensity no longer shows the
exponential decay behaviour characteristic of thermal diffusion
but rather an oscillatory relaxation, see figure 4.

4.   MASS TRANSPORT

    4.1.   Translational diffusion

    Mass diffusion by Brownian motion is an extrmely slow process.
From the Einstein equation, the time $\tau_{\text{diffusion}}$ necessary for a
molecule to diffuse over a distance 1 is given by:

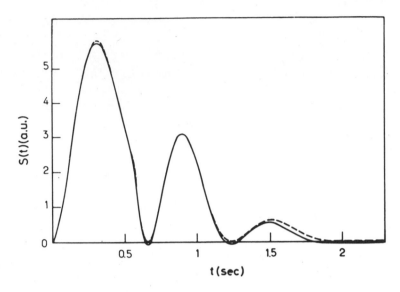

Fig. 4.   Taken from reference (23). Time dependence of the
          diffracted light intensity by a thermal grid applied to
          an overstabilized Bénard cell (i.e. with a strong inverted
          temperature gradient). The oscillatory behaviour is
          indicative of a propagating thermal mode.

$$\tau_{diff.} = \frac{l^2}{D}$$

Typical diffusion coefficients D for liquids are in the range of
$10^{-6} - 10^{-8}$ cm^2 s^{-1}. It is thus clear that the experimental time
becomes rapidly prohibitive when the diffusion takes place over
centimetric distances, as in radioactive tracers experiments.
$\tau_{diff.} = 10^8 s$ with l=1cm and D = 10^{-8} cm^2 s^{-1}, i.e. three years!
Since $\tau_{diff.}$ decreases as the square of the diffusion length,
FRS appears as an ideal candidate. Typical fringe spacings are
in the range of a few optical wavelengths and can be measured with
a high degree of accuracy by counting $\simeq$ 100 fringes within the
interference pattern. With l = 1μm, $\tau_{diff.}$ falls down by eight
orders of magnitude to 1 s.

    The experimental set-ups for mass diffusion and thermal
transport are identical. The only difference is in the nature of
the induced grating which will diffract the probe beam: thermal
grating in one case, concentration grating in the other. In
that latter case, it is not obvious to see how the partial absorption of an optical interference pattern can create a periodic
concentration distribution. The solution is based on the use
of photochromic molecules (24). The overall density of molecules

will not be changed, nevertheless a periodic distribution of
photoexcited molecules will be established within the sample. As
for the thermal grid, a phase and/or absorption grating is
associated to this distribution of photoexcited molecules since
their optical polarizability has been changed in the photochromic
process. Following the flash excitation the concentration
distribution will be gradually smeared out by diffusion. The
diffraction image will decay exponentially with a time constant
equal to $\tau_{diff.}$ ($\tau_{diff.}$ /2) in the heterodyne (homodyne) detection
scheme.

This method has been applied to various systems. Binary
mass diffusion coefficients have been obtained in the nematic
liquid crystal p-methoxy benzylidene p-n butyl aniline (M.B.B.A.)
doped with low concentrations of methyl red (MR) (25). MR is an
azo dye which undergoes a cis to trans photoisomerization with an
excited life time of several seconds (26). Since its molecular
shape is close to that of the host molecules, the binary mass
diffusion coefficient is assumed to closely approximate the true
self-diffusion coefficient D of M.B.B.A. The results for the
relaxation time of the diffraction image are shown on figure 5.
D is obtained from the slope of the curve. Nematic liquid
crystals are anisotropic materials and two different D are
measured according to the fringes orientation relative to the
macroscopic orientation axis of the sample. Both relaxation times
vary linearly with q^{-2}, insuring that we are dealing with a
diffusion process. This linear relationship stays valid as long
as the intramolecular relaxation time τ_{intra} of the excited
MR molecules is long compared to $\tau_{diff.}$ $> \tau_{intra.}$, the decay of
the diffraction image is no longer controlled by a diffusion
process and becomes q-independent. In this particular experiment,
care has also to be exercised with the reading laser beam.
It has been observed to induce a reversed photochromic effect and
to return the excited trans state to the ground cis state
prematurely (27).

True self-diffusion measurements by FRS have also been
reported in solutions of flexible macromolecular chains (28).
Prior to the experiment, the chains have been labeled chemically
at one end with a spiropyran photochromic molecule (24). For
such large objects as polymer coils of molecular weight 10^5 or
higher, the addition of a small probe of molecular weight $\simeq 300$
does not change the chain properties. The concentration of labeled
chains must however be kept low to avoid a possible segregation
of the photochromes into small micelles. In that respect,
the spiropyran is particularly well suited because its high
absorption coefficient in the excited state allow concentrations
of 10^{-8} mole/liter to be detected by FRS.

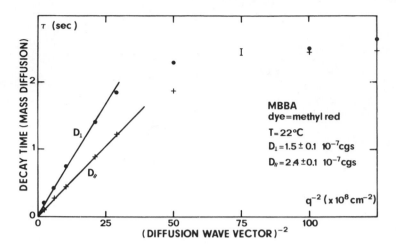

Fig. 5. Taken from reference (25). Decay time of the photo-
         chromic concentration grating vs the inverse square of
         the grating wavevector in the nematic phase of p-methoxy
         benzylidene p-n butyl aniline. Two relaxation times are
         given for the two principal molecular orientations
         relative to the fringe pattern.

         The self-diffusion coefficient of these labeled polymer
chains is largely different according to the concentration regime
as shown on figure 6. At low chain concentrations (in g/cm^3), the
chains are isolated and behave as independent particles. The
diffusion coefficient is only weakly dependent on concentration
with a slight increase as the cross-over concentration C* is
approached. C* corresponds to the concentration of first overlap
between the coils, i.e. the point where the macroscopic density
of monomers in the solution is just equal to the average monomer
density within a single coil. At high concentrations, above C*,
the chains overlap and become entangled. Following the reptation
model introduced by de Gennes (29), a given chain has to worm
its way through the maze of the other in order to diffuse.
Physically, two chains cannot cross each other, which introduces
strong topological constraints on the chain motion. The diffusion
coefficient should <u>decrease</u> with concentration as $C^{-1.75}$, as
indeed observed experimentally.

         It is worth pointing out that two distinct diffusion
coefficients can be associated to the so-called semi-dilute regime.
The first is the self-diffusion coefficient D_{self} of a few
individual labeled chains and has just been discussed in connection
with FRS. The second is a cooperative diffusion coefficient D_{coop}
for the fluctuations in the local monomer concentration. To
relax such fluctuations, the chains do not have to disentangle and

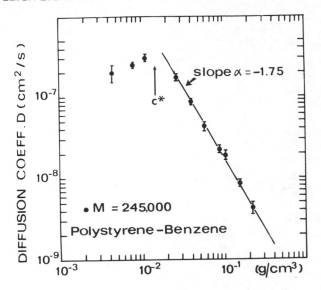

Fig. 6.   Taken from reference (28). Mass diffusion coefficient
of polystyrene chains dissolved in benzene vs polymer
concentration. C* is the concentration of first overlap
between chains of molecular weight 245,000.

they only have to move all together like a sponge. D_{coop} <u>increases</u>
with concentration as $c^{0.75}$ indicating that the elastic restoring
force to an excess of monomers in a given region of space increases
with the density of entanglements and thus with the chain
concentration. QELS experiments (30) have detected that second
diffusion mode and gave results in quantitative agreement with
the theoretical predictions. The complementary aspect of the two
light scattering techniques QELS and FRS can be seen here. FRS
is akin to inelastic incoherent neutron scattering (31) and
measures self-diffusion. QELS is comparable to inelastic coherent
neutron scattering and measures cooperative diffusion (32). The
two techniques becomes equivalent when cooperative diffusion
rejoins self-diffusion, i.e. when the diffusion particles move
independently without interactions (see the dilute regime of
figure 6).

Although photochromism has been used extensively to create
distributions of labeled molecules, it is not applicable to all
systems. In the search for new ways of inducing periodic
concentration distributions, it should be borne in mind that FRS

does not strictly require light beams but for the detection. It
is conceivable that a standing ultrasonic wave in binary solutions
will build a concentration grating if the chemical potential has
a non zero second order term in its virial expansion with pressure.
Although this is generally a weak effect, it may be detectable
in view of the sensitivity of FRS.

A few experiments have been reported with indirect couplings
between the excitation light interference pattern and the inhomo-
geneous concentration distribution. Chan and Pershan (21,15)
have measured the diffusion of water in the smectic lipid-water
phase of dipalmitoyl phosphatidyl choline (DPPC) by deforming the
lipid bilayer structure with a thermal grid. It is known that
the bilayer thickness decreases with increasing temperatures.
Thus non-uniform heating induces non-uniform thicknesses for the
lipid bilayers. The elastic pressure caused by the deformed
bilayer structure forces water to flow from regions of lower
to higher local temperature. In that case, the diffusion of
water is studied through a thermo-elastic coupling mechanism.

In binary mixtures, the Soret effect provides still another
coupling mechanism. When a temperature gradient is applied to a
binary mixture, a differential flux of species A and B sets up
and leads to a concentration gradient. Thyagarajan and Lallemand
(33) have used a carbon disulfide-ethanol mixture chosen for
its large difference between the refractive indices of the two
components. A periodic temperature pattern is produced in the
sample and the grating decay time is followed by diffraction of
a low intensity probe beam. Two different characteristic
times are observed, one short associated with the thermal
diffusivity, one long associated with the mass diffusion,
figure 7. This experiment is not easy because of the large
difference between the two mass D and thermal K diffusion
constants. As pointed out by the authors, the amplitude of the
slow decay is proportional to the ratio D/K which is extremely
small $\simeq 10^{-3}$. Consequently, long signal averaging times are
required and it is not certain that the method is practical
except in special cases.

4.2 Rotational diffusion

Polarized FRS provides an elegant way of studying the
rotational diffusion of molecules in fluid solutions. The
experimental approach is to use the pulsed optical interference
pattern to induce a distribution of preferentially oriented
molecules. For a linearly polarized probe light beam, the sample
will behave as a transient diffraction grating since in the excited
regions the optical refractive index (or the absorption) will be
different. The temporal decay of the diffracted light intensity is
a direct measure of the orientational relaxation time. This is

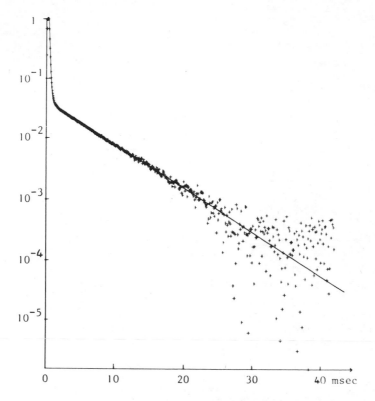

Fig. 7. Taken from reference (33). Time dependence of the
         diffracted light intensity in a carbon-disulfide-ethanol
         mixture. The data are fitted with two exponentials
         corresponding to fast thermal diffusivity and slow mass
         diffusivity driven by the Soret effect.

strictly true only if the rotational diffusion is much faster
than the translational diffusion. The two relaxation times can
however be adjusted by varying the wavevector of the pump pattern,
q_o, since the rotational time is q_o - independent while the
translational time increases q_o^2.

   Several photoselection mechanisms are possible. Using high
peak power pico-second pulses, Shank and Ippen have induced a
dichroism by saturating the absorption of molecules aligned parallel
to the pump polarization. However, they have not used FRS but
classical transmission measurements with a pump and probe set-up
(34). Phillion et al (35) have used the difference in cross-
sections for $S_o \longrightarrow S_1$ transitions according to the initial
molecular orientation relative to the pump polarization. The figure
8 shows their experimental set-up, which is typical of pico-second
laser apparatus with an optical delay line for the probe pulse.

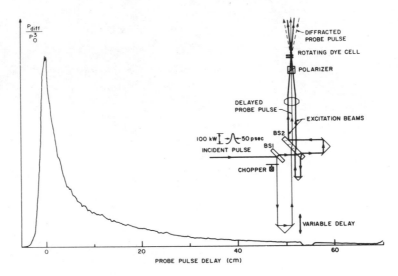

Fig. 8. Taken from reference (35). Schematic of transient grating
apparatus with pico-second pulsed excitation and
detection. A typical time dependence of the diffracted
light intensity is shown for a 10^{-4}M solution of
Rhodamine 6G in methanol.

They measure the rotational relaxation time of rhodamine 6G
in various alcohols. Typical times are ~ 200 ps. It should be
noted however that the methods based upon the excitation of
photoselected upper electronic excited states are limited to
measurements of orientational relaxation times short compared to
the intrinsic relaxation to the ground state.

There is an alternative technique based on the optical
Kerr effect. Jennings et al (36) have demonstrated that the
electric field associated with a laser beam is sufficient to induce
an optical birefringence in pure liquids and in solutions of rod-
like particles (rigid polymers, viruses, etc...). The molecules
orient through the interaction of the electromagnetic wave and
their polarizability anisotropy. No experiments with spatially
modulated light beams have been reported to-date however such a
FRS scheme should increase largely the detection sensitivity over
transmission methods. A periodic distribution of the orientation
of the molecules can also be achieved by using spatially alternating
electric fields between semi-transparent interdigital electrodes.
Such a variant of FRS has been used by Ricard and Prost (37)
to induce periodic distorsions of the molecular alignment in
smectic liquid crystals and to study the propagation velocity of
the undulation-compression mode for the smectic layers (the so-called
second sound wave).

5.   FLASH PHOTOLYSIS

Flash photolysis is the ideal technique to study the kinetics
of photochemical products by the direct measurement of their
absorption spectra (38). We will discuss here how FRS can increase
the detection sensitivity of flash photolysis to extremely small
absorption changes. The possibilities of FRS in that domain have
been recently demonstrated in two different time scales, spanning
more than fourteen decades (26,35,39). In these experiments,
the pulsed optical interference pattern at wavelength λ_o incident
on the sample is used to produce a periodic distribution of photoexcited states, e.g. cis to trans isomerization, excited singlet
states, intersystem crossing to triplet states, etc... This
transient hologram is then probed by the weak laser beam, at a
wavelength λ_1 which may or may not be different from λ_o. The
decay time of the induced grating is directly related to the
lifetime of the excited state. In the absence of mass diffusion,
the decay time for an intramolecular relaxation process of an
excited molecule to its ground state is expected to be independent
on the excitation wavevector q_o. This will be true in fluid
solutions where mass transport is bound to occur as long as q_o is
sufficiently small so that $\tau_{intramolecular} \ll \tau_{diffusion}$. A fringe
spacing of 200μm together with a diffusion coefficient $\sim 10^{-5}$ $cm^2 s^{-1}$
typical of liquids yields $\tau_{diffusion} \sim 40$ s. As this time is large,
problems with FRS will only arise for the very low-lived photochemical processes. Even so, it should be possible to use highly
viscous fluids or even organic glasses.

It has been demonstrated that FRS offers several advantages
for flash photolysis.

1.   Contrary to classical techniques which only detect absorption
      changes, FRS is sensitive to a change of either the real part
      or the imaginary part of the refractive index. This can be
      precious for preliminary investigations on materials with
      poorly known photochromic characteristics. It is clear however
      that this may in turn be an inconvenient in quantitative
      measurements. The scattered light intensity is proportional
      to the square (in the homodyne detection scheme, see section
      2) of both the optical index change and the absorption change.
      Both effects are intertwined and not easy to separate. A
      possible bypass would be to study the light diffracted into
      orders higher than I. A purely sinusoîdal amplitude grating
      diffracts only into the first order(the Fourier transform of
      a sinusoîd is a delta function) while a phase grating diffracts
      into all orders with an intensity distribution obeying Bessel
      functions (40);

2.   FRS is extremely sensitive. The limits of sensitivity have
     been studied in details by Pohl (12). With special care,
     changes in the real part of the refractive index as small as
     10^{-13} should be detectable (see table 1 of reference 12).
     Recent experiments report changes of 10^{-7} - 10^{-8} in the
     refractive index, changes of 10^{-4} - 10^{-5} o.d. in the
     absorption. The detection is facilitated by the fact that,
     unlike classical differential methods measuring transmitted
     beams, it is not necessary to detect small induced changes
     by subtraction of two much larger signals.

3.   Permanent photobleaching can be detected at the same time
     as photochromism. The figure 9 shows an example where the
     diffracted light intensity does not decay back to zero because
     of partial photodegradation.

     In a FRS experiment, the molecules in between two bright
optical fringes will never be photoexcited. This provides an
easy comparison between their ground state and the apparent ground
state to which the other molecules return after excitation to
higher states. If there is a difference at infinite times, it
means that some photodegradation has occured.

     So far we have only discussed intramolecular photochemical
processes. They are indeed the most common. It is striking however
that FRS has also permitted the first time direct observation of the

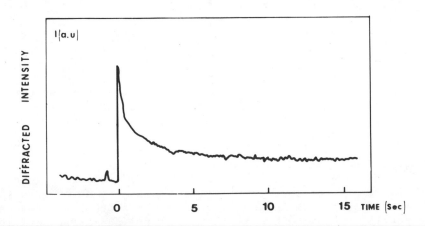

Fig. 9.   Taken from reference (26). Time dependence of the
          diffracted light intensity due to the cis-trans photo-
          isomerization of the azo dye methyl red dissolved in a
          rigid glass. The optical absorption change measured
          independently is less than 10^{-5} o.d. at the observation
          wavelength. The finite value at infinite times indicates
          a slight photodegradation.

rate of energy transport in molecular crystals (39). Contrary to the previous discussion, the decay time is no longer independent on the wavevector of the excitation grating q_o. Figure 10 shows that the transient grating decay time increases linearly with q_o^{-2}, which is typical of a diffusive process.

A diffusive energy transfer occurs between neighbouring photoexcited molecules by non-radiative intermolecular interactions. Such processes should be also observable in fluid solutions since the time scales involved for the energy transfer are always much shorter, ~ 500 ps, than those for mass diffusion.

6.   HYDRODYNAMICS

de Gennes has been the first to point out that FRS could be used to measure velocity gradients in both laminar and turbulent flows (41). This is of considerable importance in hydrodynamics because the presently available techniques such as laser anemometry measure only the local average velocities.

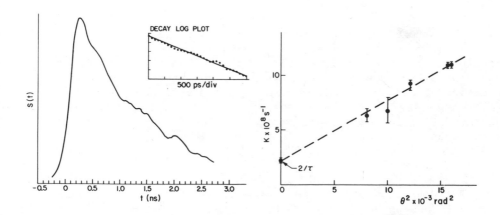

Fig. 10.   Taken from reference (39). a) Time dependence of the diffracted light intensity by photoexcited p-terphenyl dissolved in molecular crystals of pentacene; b) The decay is due to a radiationless excited-state diffusion as evidenced by the dependence of the decay rate K with the square of the crossing angle for the two exciting beam ($\theta \propto q_o$). τ is the natural excited state lifetime for an isolated molecule.

The crux of FRS is to write an optical grating in the fluid at time t = 0 and to observe its subsequent distorsions as it is sheared and strained by the local velocity gradients. In a laminar flow, a velocity gradient $\bar{\bar{G}}$ ($G_{ij} = \partial V_i / \partial x_j$) induces a rotation and a shearing of the grating. If $\vec{q_o}$ is the wavevector of the initial grating at time t = 0 and $\vec{q_i}$ and $\vec{q_D}$ those of the incident and diffracted probing laser beam, the relationship between $\vec{q_i}$, $\vec{q_D}$, $\vec{q_o}$ and $\bar{\bar{G}}$ writes:

$$\vec{q_D} = \vec{q_i} + \vec{q_o}(1 + \bar{\bar{G}}t)$$

This equation is valid only if Gt << 1. On the contrary a spatially uniform velocity field (no velocity gradient) will leave the wavevector of the diffracted beam unchanged. The average velocity could however be detected in the heterodyne detection scheme as a periodic time modulation of the diffracted light intensity as the relative phase of the optical grating to the static local oscillator will be continuously changing (modulo 2π)

$$V_a(t) \propto |B| |E_s(t)| \cos \emptyset$$

In a turbulent flow the equations are more complicated and require the calculation of the so-called Richardson functions which describes the pair correlation function between two distant particles. The application of FRS to this problem has recently been discussed by Petit and Guyon (42).

So far the only hydrodynamic experiment using FRS has been performed in a plane Poiseuille flow (43). The printed optical grating is a thermal grid of .5°K amplitude and 54 µm fringe spacing. The diffraction image of the grating is continuously monitored as a function of elapsed time. Figure 11 a and b, taken from reference (43) show two photographs of the diffraction dots a few nano-seconds after the Q-switched excitation laser pulse and 600 microseconds respectively. A rotation of the image is clearly observed. Here, $\beta_2 - \beta_1 \simeq 10°$, which yields a velocity gradient of 300 s^{-1}.

The variation of the probe beam along an axis perpendicular to the flow pipe allows to determine the complete velocity gradient field. As shown on figure 12, the velocity gradients are maxima close to the pipe walls and zero in the center. This is in full agreement with the parabolic velocity profile of a Poiseuille flow. The curves correspond to two different average velocities of 37 cm s^{-1} and 20 cm s^{-1} (respective flow rates are 74 cm^3 s^{-1} and 41 cm^3 s^{-1}). The spatial resolution is limited by the size of the probing beam to about 300µm. The data accuracy on velocity gradients is mainly limited by the angular resolution of the β angles. It could be improved by allowing for longer observation times since β increases linearly with time ($\beta = \bar{\bar{G}}t$). In that respect,

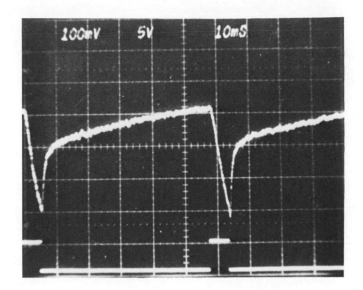

Fig. 11. Taken from reference (43). Photographs of the diffraction pattern a) a few nanoseconds after the exciting interference pattern b) 600 microseconds later. The image has rotated by an angle $\beta = \beta_1 - \beta_2 = 10^\circ$ for $\frac{\partial V}{\partial x} \sim 300 \ s^{-1}$.

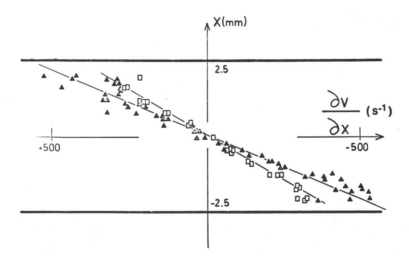

Fig. 12. Taken from reference (43). Measurement of the velocity gradient across the flow cell for two different flow rates, ▲ 74 $cm^3 \ s^{-1}$ and ▢ 41 $cm^3 \ s^{-1}$, in the laminar Poiseuille regime.

a thermal grid is not ideal since the decay time of the grating by thermal diffusivity alone is short, $\sim 200\mu s$ (see section 3). A large improvement would be to use a concentration grating of photoexcited molecules as described in section 4. The natural decay time by mass diffusion is typically 3 orders of magnitude longer than by thermal diffusion.

Measurements in the turbulent regime of the Poiseuille flow are currently under way. There is no doubt that it will be a rapidly expanidng domain of research. The possibility of studying a wide range of eddies scales (few microns - millimeters) by changing the grating period makes FRS very attractive for turbulence.

Another domain accessible to FRS is the externally-driven hydrodynamic instabilities. A recent example is given by the pretransitional phenomena in the homogeneous regime preceding the onset of the Rayleigh-Benard instability. When a horizontal fluid layer is heated from below, circular rolls develop when the temperature gradient across the cell exceeds a critical value. Below threshold, no macroscopic fluid motion can develop but hydrodynamic modes continue to be continuously excited. In analogy with equilibrium phase transitions, theory predicts that the amplitude of these thermally-excited fluctuations should diverge at threshold. This critical slowing down should be observable by QELS. However, the low wavelength of the first unstable mode, imposes to work in the forward direction, with all the associated stray light problems, and the experiment has only been achieved very recently with FRS (44). The results corroborate the theoretical predictions. The decay time of the thermal fluctuations externally induced by an optical interference pattern heating a slightly absorbing isotropic fluid increases by a factor of five when the instability threshold is approached from below. Moreover, the effect is more pronounced when the thermal grating periodicity is equal to the optimum value of twice the cell height. One can imagine similar experiments for the Couette instability between two rotating vertical cylinders (45).

CONCLUSION

FRS falls into the well-investigated category of the fluctuation spectroscopy techniques. If one tries to explain why it has been applied so successfully to many fields of research, the main answer is that the fluctuations are excitable well above their statistical thermodynamic values. The signal to noise ratio of the method is then extremely good. Furthermore the spatial properties of the induced gratings concentrate the scattered light intensity into a narrow solid angle, minimizing stray light problems. The absence of signal in the absence of external excitation makes the method truly differential. i) The amount of extraneous light can always be precisely measured during the blank periods and later

subtracted. ii) Small useful signals are not buried in large
backgrounds. Since these signals are obtained with pulsed
excitation beams, the detection electronics is particularly
simple. Indeed FRS directly yields the time dependence $\Delta S(q_o, t)$
of the measured quantity and not its correlation function
$< \Delta S(q_o, t) \; \Delta S(q_o, t + \tau) >$ All these advantages make FRS competitive
with the best interferometric techniques. Refractive index
changes of 10^{-11} and/or absorption changes of less than 10^{-5} o.d.
have already been detected. For a liquid, this means that a
temperature modulation of less than 10^{-7} ^oK is enough to produce a
detectable change and to measure the thermal transport properties
for instance. The consequence of this extreme sensitivity is that
the induced perturbation will generally not perturb the system under
study. In phase transition studies, the main limitation to going
close to the transition temperature will be set by the temperature
controllers and not by the FRS. Similarly, the threshold for
hydrodynamic convective instabilities will not be affected. When
dopants have to be added, their concentration can be kept to a
level corresponding to detectable trace impurities. Experiments have
been performed with concentrations of photochromix molecules less
than 10^{-8} mole/liter, which puts FRS in the same performance range
as fluorescence.

The possibility of working with extremely small sample
volumes should be underlined. An extreme case are the true
mouolayers at liquid-liquid or liquid-air interfaces which should
be amenable to investigation by FRS. Limited sample volumes are
also useful with scarce biological products or anisotropic
materials difficult to align over large scales. Among many examples,
one can cite the increasingly popular thermotropic and lyotropic
liquid crystalline systems (nematic polymers, lipid-water multi-
layers, ...). Otherwise isotropic systems can also become anisotropic when the space dimensionality is reduced. A flexible polymer
chain becomes elongated when it is trapped in a long cylindrical
pore of diameter smaller than the coil radius of gyration. Interactions with solid surfaces can induce an order in the adjacent
fluid layers.

In view of all its possibilities, FRS appears as an important
complement to the existing optical techniques. There is no doubt
that, in the near future, FRS will be applied to a growing number
of research fields in chemistry, physics and biology.

ACKNOWLEDGEMENTS

The author gratefully acknowledges many helpful discussions
with H. Hervet, L. Léger, W. Urbach, P. Lallemand and D. W. Pohl.
This work was partially supported by the CNRS under ATP no.3455 and
by the DGRST under contract no.77-7-1459. The laboratory is
associated to the Centre National de la Recherche Scientifique ERA
no.542.

References

1. H. Z. Cummins and H. L. Swinney in Progress in optics. Vol.8. E. Wolf, ed., Wiley, New York (1970).
2. G. B. Benedek. In Polarisation, Matière et Rayonnement, Livre de jubilé en l'honneur du professeur A. Kastler, Presses Universitaires de France, Paris (1968).
3. N. C. Ford. Chem. Script., 2:193 (1972).
4. J. B. Lastovka and G. B. Benedek, Phys. Rev. Lett. 17:1039 (1966).
5. B. Chu, Phys. Rev. Lett., 18:200 (1967).
6. S. B. Dubin, J. H. Lunacek and G. B. Benedek, Proc. Nat. Acad. Sci. U.S., 57:1164 (1967).
7. G. Durand, L. Léger, F. Rondelez and M. Veyssié, Phys. Rev. Lett., 22:1361 (1969).
8. R. I. Scarlet, Phys. Rev. A6:2281 (1972).
9. H. Eichler, G. Salje and H. Stahl, J.Appl. Phys., 44:5383 (1973).
10. D. W. Pohl, S. E. Schwarz and V. Irniger, Phys. Rev. Lett., 31:32 (1973).
11. N. Bloembergen, W. H. Lowdermilk and C. S. Wang, Phys. Rev. Lett., 25:1476 (1970).
12. D. W. Pohl, IBM J. Res. Develop., 23:604 (1979).
13. L. D. Landau and E. M. Lifshitz, Statistical physics, Pergamon, London (1958).
14. A. M. Glass. In: Photonics, M. Balkanski, ed., Gauthier-Villars, Paris (1975).
15. W. K. Chan and P. S. Pershan, Biophys. J., 23:427 (1978).
16. P. Calmettes and C. Laj, J. Physique, 33:C1-125 (1972).
17. W. Urbach, H. Hervet and P. Rondelez, Mol. Cryst. and Liq. Cryst. 46:209 (1978).
18. F. Rondelez, H. Hervet and W. Urbach, Phys. Rev. Lett., 41:1058 (1978).
19. W. Urbach, H. Hervet and F. Rondelez, to be published (1980).
20. D. W. Pohl, this book.
21. W. K. Chan and P. S. Pershan, Phys. Rev. Lett., 39:1368 (1977).
22. J. A. Cowen, C. Allain and P. Lallemand, J. Physique Lett. 37:L-313 (1976), C. Allain and P. Lallemand, J. Physique, 40:693 (1979).
23. J. P. Boon, C. Allain and P. Lallemand, Phys. Rev. Lett. 43:199 (1979).
24. See e.g., G. H. Brown. Photochromism - Techniques of chemistry vol. 3, Wiley, New York (1971).
25. H. Hervet, W. Urbach and F. Rondelez, J. Chem. Phys., $6\frac{3}{4}$:2725 (1978).
26. F. Rondelez, H. Hervet and W. Urbach, Chem. Phys. Lett., 53:158 (1978).
27. W. Urbach, private communication.

28. H. Hervet, L. Legér and F. Rondelez, Phys. Rev. Lett. 42:1681 (1979).

29. P. G. de Gennes, Macromolecules, 9:594 (1976).

30. M. Adam and M. Delsanti, Macromolecules, 10:1229 (1977).

31 See, e.g. A. Maconnachie and R. W. Richards, Polymer 19:739 (1978).

32. It was recently pointed out by M. Weissman that self-diffusion measurements by QELS are possible even in strongly interacting systems, provided there is some distribution of the scattering amplitudes (e.g. via polydispersity for polymer solutions).

33. K. Thyagaraja and P. Lallemand, Optics Com., 26:54 (1978).

34. C. V. Shank and E. P. Ippen, Appl. Phys. Lett., 26:62 (1975).

35. D. W. Phillion, D. J. Kuizenga and A. E. Siegman, Appl. Phys. Lett., 27:85 (1975).

36. B. R. Jennings and H. J. Coles, Proc. R. Soc. Lond. A348:525 (1976).

37. L. Ricard and J. Prost, J. Physique, 40:C3-83 (1979).

38. N. J. Turro, Molecular photochemistry, Benjamin, New York (1967).

39. J. R. Salcedo, A. E. Siegman, D. D. Blott and M. D. Fayer, Phys. Rev. Lett., 41:131 (1978).

40. K. S. Pennington, in Handbook of lasers, R. J. Pressley ed., CRC Press, Ohio (1971).

41. P. G. de Gennes, J. Physique Lett., 38:L1 (1977).

42. L. Petit and E. Guyon, J. Physique, in press (1979/80).

43. M. Fermigier, E. Guyon, P. Jennfer and L. Petit, Preprint (1979).

44. C. Allain, H. Z. Cummins and P. Lallemand, J. Physique Lett. 39:L474 (1978).

45. G. I. Taylor, Phil. Trans. Roy. Soc. (London) A223:289 (1923).

46. Copyright 1979. American Institute of Physics New York; reprinted with permission from reference 23.

47. Copyright 1978. North Holland Publishing Co., Amsterdam; reprinted with permission from reference 33.

48. Copyright 1975. American Institute of Physics New York; reprinted with permission from reference 35.

49. Copyright 1978. American Institute of Physics New York; reprinted with permission from reference 39.

50. Copyright 1979. American Institute of Physics New York; reprinted with permission from reference 43.

STUDY BY FORCED RAYLEIGH SCATTERING OF THERMAL EXCITATIONS IN

A STRATIFIED LIQUID LAYER

C. Allain, P. Lallemand and J. P. Boon*
Laboratoire de Spectroscopie Hertzienne de l'ENS
Paris

*Service Chimie Physique – Université libre de Bruxelles
Bruxelles

INTRODUCTION

We present a brief review of the work done at Ecole Normale on the thermal excitations of a horizontal liquid layer in the presence of a vertical temperature gradient. This old problem has been studied by many authors especially in cases where the temperature gradient is large enough to lead to the Rayleigh-Benard instability (1). The recent renewed interest for these studies has benefited from the use of new techniques like laser Doppler velocimetry (2). Here we shall consider only situations in which there is no spontaneous macroscopic motion of the fluid of large amplitude. This occurs when the temperature gradient is stabilizing (hot plate above in the case of fluid of positive thermal expansion coefficient), or when it is not large enough to destabilize the layer. In a first part we shall recall a few theoretical results as a guide line for the experimental studies. These will be described in the second part, making use of the very efficient forced Rayleigh scattering.

THEORETICAL DISCUSSION

Let us consider a horizontal fluid layer of thickness d, placed in between two rigid plates, of very large thermal conductivity, that are maintained at different temperatures T_o and $T_o + \beta d$. The relevant fluid properties are its thermal diffusivity K, kinematic viscosity ν and thermal expansion coefficient α. We shall use later the Prandtl number $P=\nu/K$. Here we are interested in a two dimensional analysis of the low frequency normal modes of the fluid..

a)   Equations of motion

Following the text book of Chandrasekhar, (3) one can start
from the linearized hydrodynamics equations to write the equations
of motion of the fluid.   If we call $\theta(x,z,t)$ and $W(x,z,t)$ the
fluctuations in the temperature and in the vertical component of
the velocity, we have for times long compared to the period of
the relevant sound waves:

$$\frac{\partial \theta}{\partial t} = \beta W + K \nabla^2 \theta$$

$$\frac{\partial}{\partial t} \nabla^2 W = \nu\nabla^4 W + g\alpha \frac{\partial^2 \theta}{\partial x^2}$$

Here x and z represent the horizontal and vertical coordinates
respectively, g is the acceleration of gravity.

To solve this set of equations, we need to include boundary
conditions.   These are

$$\left. \begin{array}{rcl} \theta &=& 0 \\ W &=& 0 \\ \frac{\partial W}{\partial z} &=& 0 \end{array} \right\} \qquad \text{for } z = \pm\frac{1}{2}$$

for rigid and perfectly conducting plates.   Note that if the
surface of the liquid were stress-free the last boundary condition
would be replaced by:

$$\frac{\partial^2 W}{\partial z^2} = 0$$

As this leads to simple results, we shall first discuss the
stress-free case and then present some results for the rigid
case.

b)   Solution of the stress-free case:

As the experimental studies to be discussed later allow us to
select a given wavelength L for the mode structure along the
horizontal direction, we shall look for solutions of the form

$$e^{st} \; e^{i2\pi \frac{x}{L}} \; f(z)$$

To satisfy the boundary conditions for W, we can choose $f(z)$
as $\cos (2m+1)\pi\frac{z}{2d}$ or $\sin 2m\pi\frac{z}{2d}$.

If we consider only the lowest order symmetric mode, the equations of motion lead to the following dispersion equ

$$(s + s_t) (s + s_v) - s_v s_t \frac{R}{R_c(a)} = 0$$

where

$s_t = Kq^2$ relaxation rate of the uncoupled temperature mode

$s_v = \nu q^2$ relaxation rate of the uncoupled "vortex" mode

$R = \alpha\beta \frac{d^4 g}{\nu K}$ Rayleigh number

$a = \frac{2\pi d}{L}$ reduced horizontal wave number

$q = \{(\frac{2\pi}{L})^2 + (\frac{\pi}{d})^2\}^{1/2}$ wave vector of the lowest even mode

$R_c(a)$ critical Rayleigh number for the mode of wavelength L.

The dispersion equation has two roots s_1 and s_2. One of them goes to 0 as $-\varepsilon\{(s_v s_t)/(s_v + s_t)\}$ goes to zero (where $\varepsilon = (R_c(a)-R)/R_c(a))$, that is when the destabilizing temperature gradient reaches the point where the fluctuations of wavelength L develop into a macro-scopic collective motion. One thus gets a critical slowing down for the wavelength L that corresponds to the lowest value of $R_c(a)$. For other wavelengths a similar slowing down takes place but one cannot reach $\varepsilon=0$ because convection has taken place.

If one applies a stabilizing temperature gradient, one finds that the dispersion equation will exhibit complex roots for $R<R_k$ with $R_k = -R_c(a) \frac{(s_v - s_t)^2}{4 s_v s_t}$. This means that the coupling of temperature and "vortex" modes leads to propagating modes, that we call thermoconvective waves. These waves propagate with a phase velocity:

$$v = \{s_v s_t \frac{R_k - R}{R_c(a)}\}^{1/2} \frac{L}{2\pi}$$

and they are damped with a damping rate

$$\frac{1}{2} (s_v + s_t)$$

c) Solution for rigid boundaries

In practice the boundaries are rigid, so that we have to
solve the problem with $\partial w/\partial z = 0$ for $z = \pm 1/2$. This was done
using either a characteristic value equation (5) or a variational
method. The details will be published elsewhere (6). We have
found that we get a very satisfactory account of all our results
using the simple dispersion equation 1, provided we replace the
kinematic viscosity ν by $\nu f(a)$, where the coefficient $f(a)$
depends only upon the reduced wave vector a, and the nature of the
mode (order m and symmetry). It is interesting to emphasize
the fact that the correction factor $f(a)$ does not depend upon the
Prandtl number of the fluid. We give elsewhere (6) the general
formulae to calculate $f(a)$.

We can thus extend the preceding remarks about the behaviour
of the relaxation rates. To a very good accuracy one root is

$$s \sim - s_t \; \frac{pf(a)}{1 + pf(a)} \; \varepsilon$$

near the convection instability. In addition s is complex for
$R < R_k$, with

$$R_k = - R_c(a) \; \frac{(1 - pf(a))^2}{4p \; f(a)}$$

and the values deduced from equation 1 are usually accurate to
better than 1%. This simple result allows a great simplification
in the planning of the experimental work. Note that when a is
very large, $f(a) \sim 1$, which means that the boundary conditions
have no significant effect. Furthermore, for the wavelength for
which the convective rolls appear (a = 3.117), $f \sim 1.93$. This
means that for fluids of low Prandtl number (in a gas $p \sim 0.7$) the
low mode of wave length L_c is the thermal mode and not the "vortex"
mode as would be the case in free space.

EXPERIMENTAL RESULTS

We shall describe now some of the experimental work that we
have done to verify the points discussed in part I. We have
used a light scattering technique which allows us to analyse
the temporal behaviour of an excitation of a given wavelength.
To choose the thickness d of the liquid cell we had to make the
following compromise: a thin cell is preferable from the optical
point of view as it leads to smaller wavelengths, but a thick
cell allows to reach larger Rayleigh numbers for a given temperature
difference. We finally took thicknesses from 1.6 to 5 mm, so
that the relevant wavelengths were of the order of 3mm. If we
want to study the corresponding fluctuations using spontaneous

light scattering, we shall work with such a small scattering
angle (of the order of 1 minute of arc) that stray light will
cause overwhelming difficulties. This is why we decide to use
the forced Rayleigh light scattering technique that allows one to
increase the signals by a very large factor and thus essentially
avoid the problems due to stray light.

a) Experimental Setup

We show in figure 1 a schematic diagram of the setup that
we used. It involves two laser beams. One of high intensity is
used to excite the mode of wavelength L, the other is used to
monitor the resulting changes in index of refraction. The heating
beam is slightly focussed to provide a uniform heating along the
vertical direction and it traverses first a grid to select one
particular wavelength L. The intensity of that beam varies in
time either sinusoidally (at frequency ω) or in pulses of duration
Θ. As the liquids we use are transparent for the argon ion laser
lines, we add to them a small amount of impurity : iodine, red
dye or aniline. We verified in each case that the critical
temperature gradient for appearance of the convective rolls is not
significantly altered by the addition of the impurity. One
may note that we do not use the same method to produce a periodic
illumination of the liquid as in the usual setup.   (7) Instead
of using an interference phenomenon, we just passed the beam
through a grid. This has several advantages:

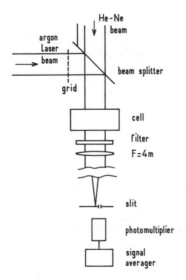

Fig. 1.   Experimental setup   for forced Rayleigh scattering
          experiments at long wavelengths.

- one can use all the lines of the argon laser,
- one can easily measure the horizontal wavelength L
- this has a very good stability, so that we are able to set at will the phase of the signal when we work in the heterodyne mode. This last feature is quite convenient as the signal recorded by the phototube is proportional to:

$$S(t) \quad = \quad |\delta \cos \phi + E_S(t)|^2$$

where δ and ϕ are the amplitude and phase of the part of the stray light that is coherent with the scattered light at the detector location.

The fact that such a grid does not lead to strict spatial sinusoidal variation of the intensity is not troublesome as we separate the various spatial harmonics (usually only the third harmonic gives a sizeable diffraction).

To monitor the corresponding index of refraction grating, we used a low power helium–neon laser (wavelength λ, $P \simeq 3\text{mW}$). If δT is the amplitude of the temperature variation due to heating, we have $\delta n = (\partial n/\partial T)p \, \delta T$ for the amplitude of the index of refraction modulation. To first order, the amplitude of the diffracted beam $E_S(t)$ will be proportional to

$$J_1(2\pi \frac{\delta nA}{\lambda})$$

where A is the optical path in the cell, and J_1 is a Bessel function of first order. To avoid any non linearity in the relationship between $E_S(t)$ and $T(t)$ we have to keep the argument of J_1 small enough. This condition leads to severe limitations in terms of the maximum heat that can be absorbed. For instance in ethanol, for A=12 mm, 10 mK leads to $2\pi\delta nA/\lambda = .52$, which is already too much.

b) Operating mode

Instead of operating the experiment in the same way as in the first reports on forced Rayleigh scattering, where the duration of the exciting beam was very short, we found it convenient to use long pulses or a modulated intensity. We shall now indicate what kind of signals will be observed in various situations, when the normal modes decay as $\cos \beta t \exp(-\alpha t)$.

i)   Pulses of duration Θ

Let us apply a heating beam of constant intensity, lasting a time, Θ; that is

$$I_H = 0 \qquad t < 0$$

$$= I \qquad 0 < t < \Theta$$

$$= 0 \qquad t > \Theta$$

Using a suitable convolution product, we get:

$$E_S(t) \propto 0 \qquad\qquad t < 0$$

$$\propto \frac{\alpha + e^{-\alpha t}(\beta \sin \beta t - \alpha \cos \beta t)}{\alpha^2 + \beta^2} \qquad 0 < t < \Theta \qquad\qquad (2)$$

$$\propto \frac{e^{-\alpha t}}{\alpha^2 + \beta^2} \{ V \cos \beta t + V \sin \beta t \} \quad t > \Theta$$

with

$$U = \{ - \alpha(1 - e^{\alpha \Theta} \cos \beta \Theta + \beta e^{\alpha \Theta} \sin \beta \Theta \}$$

$$V = \{ \alpha e^{\alpha \Theta} \sin \beta \Theta + \beta(1 - e^{\alpha \Theta} \cos \beta \Theta \}$$

Using this form, we find that the signal S(t) reaches its first maximum for $t = \pi/2\beta$ whatever the values of the local oscillator (δ and ϕ). We shall use later this property to determine β.

ii) Sinusoidal excitation

Let us now apply $I_H = I \cos \omega t$

we find:

$$E_S(t) \propto \frac{\alpha}{\alpha^2 + \beta^2} + \frac{1}{2} \cos \omega t \{ \frac{\alpha}{\alpha^2 + (\omega + \beta)^2} + \frac{\alpha}{\alpha^2 + (\omega - \beta)^2} \}$$

$$+ \frac{1}{2} \sin \omega t \{ \frac{\omega + \beta}{\alpha^2 + (\omega + \beta)^2} + \frac{\omega - \beta}{\alpha^2 + (\omega - \beta)^2} \}$$

We obtain an oscillating contribution whose amplitude and phase is given by:

$$A \propto \{ \frac{\alpha^2 + \omega^2}{(\alpha^2 + \beta^2 - \omega^2)^2 + \cdot 4\alpha^2 \omega^2} \}^{1/2}$$

and:

$$\Phi = \text{Arctg} \; \frac{\omega(\alpha^2-\beta^2+\omega^2)}{\alpha(\alpha^2+\beta^2+\omega^2)}$$

c) Experimental results:

i) Approach to convection:

We have first studied the approach to convection (8). For that purpose, we used a cell thickness d=1.65 mm filled with ethanol (p=16.6). The observed temperature difference ($\Delta T = 5.25$ ºC) for appearance of convective rolls agrees well with the calculated value (corresponding R_c=1707). We then measured the decay rate $|s|$ of the slow mode for various wavelengths L and temperature differences. From $|s|$ we deduced an effective thermal diffusivity $\Gamma^* = |s| / q^2$, which is plotted in figure 2. When no gradient is applied, we obtain a value close to the known value of K (9.34×10^{-4} cm^2/sec) in ethanol. For $\Delta T \neq 0$, we get a qualitative agreement with the calculated curves. We shall discuss later a possible explanation for the scatter of the experimental points.

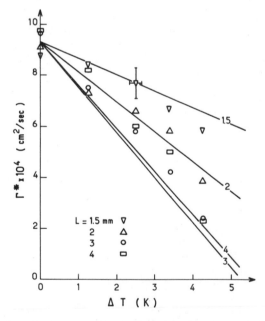

Fig. 2, Effective thermal diffusivity of ethanol in a 1.65 mm thick layer vs vertical temperature differences, as determined by forced Rayleigh scattering for various values of the wavelength L. The solid lines correspond to the slow modes derived from the dispersion equation.

ii) Thermoconvective Waves:

We have performed experients (9) in acetone which has a
fairly small Prandtl number (P = 3.6). We first used the
sinusoidal excitation to measure the amplitude and phase of the
scattered electric field. We. show in figure 3 the results obtained
for d=2.5 mm, L = 3mm and a Rayleigh number of −69000. Note
that in this case, a = 6.074, R_c(a) = 3527, f(a) = 1.203,
R_k = −2360. The calculated values of α and β are 1.67 and
5.61 s^{-1} respectively. The solid line is drawn for α = 1.88 s^{-1}
and β = 5.34 s^{-1} which gave the best fit to the data. Starting
from these experimental values, we calculated the phase which
is compared to its measured values in figure 4. We thus see
that the model of part I allows a quite satisfactory description
.of the experimental results.

Similar measurements have been performed for a series of
experimental situations and were all in satisfactory agreement
with theory. Now it turns out that extracting the amplitude
from the recorded signals. is somewhat delicate due to the presence
of some stray light which may not be constant during a whole run.
We thus decided to use mostly data recorded in the pulse mode.

Analyzing the temporal behaviour of the recordings, we
found that they could be fitted with formulae derived from
equation 2. This gave us confidence into considering only the
heating part of the experimental cycle and in particular the
instant t_m for which the signal is maximum. This allowed us to

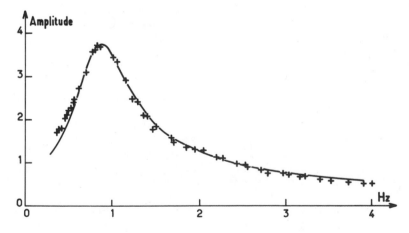

Fig. 3. Amplitude of the thermal response of a fluid layer
         vs frequency of the excitation.

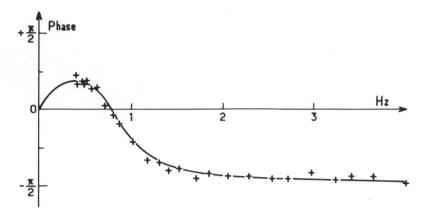

Fig. 4.   Phase of the thermal response of a fluid layer vs
          frequency of the excitation.

determine simply $\beta(\equiv\pi/2t_m)$. We thus plotted β^2 for several
values of $|R|$. We show in figure 5 an example of such a plot
for a cell of thickness d = 2.5 mm and a wavelength L=3 mm.
The theory of part I leads to the solid curves. Here "0"
corresponds to the first even mode, "1" to the first odd mode
and "2" to the second even mode. We systematically found
experimental values that were in between curves "0" and "1".
We thus get only a qualitative agreement between theory and
experiment, but we think that this is probably due to some
difficulties in the experimental procedure as we discuss below.

     When we apply the steady temperature gradient that couples
the temperature and "vortex" modes, the liquid behaves like a
prism and the light beams are bent while propagating. As a
result the heating may not be homogeneous along the vertical
direction, and thus we may have excited not only the mode "0"
but also higher order modes. Now in addition to this prism
effect, the cell plays the role of a narrow horizontal slit.
In the absence of temperature gradient, we observe on the
detecting beam a very strong diffraction effect giving rise at
the detector to a symmetric vertical pattern. When the
temperature gradient is applied, this pattern not only moves
(due to the prism effect) but becomes severely distorted in an
asymmetric way. As a result it was not possible to separate
at the detector location the patterns due to the various modes.

     The signal that we recorded was thus a mixture of pure
signals due to the various modes, but with unknown relative
weights. We found that for given conditions (d, L, ΔT) we
could record various temporal dependences by choosing different
parts along the vertical diffraction pattern, or by modifying the

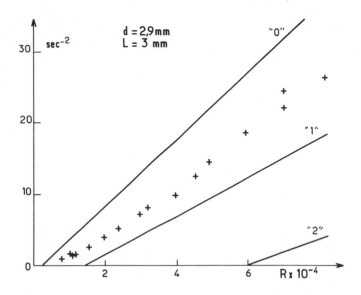

Fig. 5.   Square of experimental value of β^2, vs Rayleigh number
in a fluid layer.

vertical distribution of the heating beam. The data in figure 5
refer to the data recorded near the central diffraction spot.

Although we did not make a quantitative study of these
effects, we believe that they are most certainly responsible for
the lack of agreement between theory and experiment. These
effects obviously appear for both signs of temperature gradient.
For instance we may very well have given too much emphasis to the
faster time variation of the signal due to mode "1" when we
determined Γ^* plotted in figure 2.

In conclusion we can summarize the content of this communication.
We have shown that the simple results concerning the normal
modes of a stratified stress-free fluid layer can be used in
the realistic rigid boundary case by a simple change in the
kinematic viscosity. We then showed that forced Rayleigh light
scattering can be usefully extended to various excitation schemes.
Finally we present experimental results that are in qualitative
agreement with the theory.

To improve the quality of the test of the theory, we plan to
perform new experiments in which better optical techniques will
allow us to avoid the difficulties that we have met here.
Replacing the global examination of the thermal grating, as we did
here, by a more detailed study using (10) differential inter-
ferometry, as used by Dubois et al (10), should allow us to make

progresses. Another advantage of that technique is that one can work with larger index gradients and so one could study non linear effects without being very close to the convective instability.

REFERENCES

1. A recent review of the literature is given by C. Normand, Y. Pomeau and M. G. Velarde. Rev. Mod. Phys., 49:581 (1977).
2. P. Berge, M. Dubois, Phys. Rev. Letters, 32:1041 (1974).
3. S. Chandrasekhar, "Hydrodynamic and hydromagnetic stability" Clarendon Press, Oxford (1961).
4. H. N. W. Lekkerkerker and J. P. Boon, Phys. Rev., A10:1355 (1974).
5. G. Z. Gershuni and E. M. Zhukhovitskii, "Convective Stability of Incompressible Fluids", Keter Publishing House, Jerusalem (1976).
6. P. Lallemand and C. Allain, submitted to J. de Physique.
7. D. W. Pohl, S. E. Schwarz and V. Irniger, Phys. Rev. Letters, 31:32 (1973).
8. C. Allain, H. Z. Cummins and P. Lallemand, J. Phys. Letters, 39:L-473 (1978).
9. J. P. Boon, C. Allain and P. Lallemand, Phys. Rev. Letters, 43:199 (1979).
10. M. Dubois and P. Berge, Private communication.

FORCED RAYLEIGH SCATTERING AS OPTOELASTIC PROBE IN S_A LIQUID CRYSTALS

F. Scudieri

Istituto di Fisica-Facolta' di Ingegneria
Universita' di Roma
Roma, Italy

Thermal and mass transport coefficients can be experimentally measured in a very simple and elegant way by utilizing the forced Rayleigh scattering technique. Also the materials, that exhibit a poor light absorption in the visible spectral region, can be studied with low energy pulses if just a little amount of absorbing dye is added to the sample. In the theory of the forced Rayleigh scattering, depending on the type of diffusion process that has to be analysed, only the related transport coefficients are involved without any characterization of the elastic properties of the sample. The thermoelastic coefficients are taken into account only in the case where tunable ultrasound trains must be produced by using the optical absorption of high intensity light pulses. A very interesting case, where also with low energy it is possible to observe elastooptic effects, is exhibited by smectic A materials.

If a homeotropically aligned smectic A sample is stressed by a negative pressure applied across the layers, due to the circumstance that the material keeps a constant layer thickness, the structure reacts by an undulation whose q vector in the layers plane is related to the attenuation depth L in the thickness direction by

$$L = 1/(q^2 \lambda)$$

and with a critical value

$$q_c = (\pi/d\lambda)^{\frac{1}{2}}$$

where λ is the penetration length of de Gennes (1,2,3) and d is

the sample thickness : such a mechanooptic effect can take place
only for a dilation greater than a threshold value δ_{th} = $2\pi\lambda$. For
very large dilations the undulation deformation degenerates into
a more complicated one characterized by a crossed grating pattern.
An effect, similar to the previous one, is observed by locally
heating the homeotropic smectic A sample: e.g. by means of an
intense light beam. In this case after heating the layered
structure is deformed: undulations and square pattern are shown as
a consequence of the thermal gradient (4). The connection between
the distortion and thermal gradients has been also emphasized
in hot wire experiments (5). A very direct check of the localized
defects, in the form a dislocations, created by means of a periodic
thermal pattern has been performed by using the acoustic emission
connected with the annealing of the defects (6).

    Due to the high birefringence exhibited by the liquid crystals
all distortions of the director field can be easily observed by an
optical probe: in this way the consequence of the previous
considerations is that the forced Rayleigh scattering technique
can be usefull to study the time behaviour of the deformations in
a smectic A sample. To test such a possibility some preliminary
observations have been performed utilizing a 1.2 W argon laser
(λ_1 = 5145 Å) as the exciting light pulse to induce the thermal
periodic pattern into a 100 µm thick sample of a room temperature
smectic A material (COB) doped with methyl red dye (5.10^{-3} M) and
homeotropically aligned. A He-Ne laser beam (λ_2 = 6328 Å) has been
utilized as the analyzing light. The scattered intensity is
detected by a photomultiplier with an input circular aperture
(diameter: 200 µm) (see fig. 1). By increasing the time duration of
the exciting laser pulse some different time behaviours for the
scattered light can be observed. If the exciting pulse energy is
low or, that it is the same at our fixed power level, the time
duration of the pulse is short (typically until 1.5 ms) the time
decay of the scattered light intensity is an exponential one
characterized by a single time constant t_1 (few ms at small angles)
connected to the diffusion process of the thermal excited grating.
By increasing the pulse time duration up to 5 ms the light intensity
is characterized by the presence of two time decay processes
(see fig. 2): the fast one is connected to the thermal relaxation
of the local refractive index change, the slow decay time τ_2 (of

the order of few tens of ms) cannot be attributed to mass diffusion
phenomena whose characteristic time is in the range between
1÷3s depending on the used spatial frequency of the temperature
grating. If we analyse a virgin sample, previously never exposed
to a light fringe pattern, at successive exciting pulses fo 2.2 mJ
(whose only few percent is absorbed by the sample) the forced
Rayleigh scattered component exhibits the time behaviour shown in
fig. 3: the light intensity connected with the slower process
increases for successive pulses until it masks completely the fast

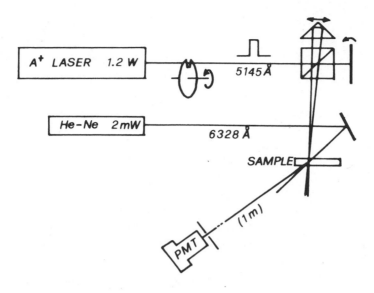

Fig. 1.   Experimental set up

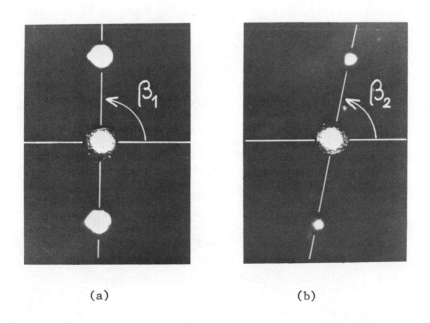

(a)   (b)

Fig. 2.   Time behaviour of scattered light.

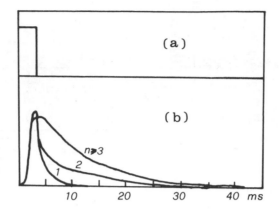

Fig. 3.   (a) Excitation light pulse (b) time behaviour for
          successive pulses in a virgin sample.

thermal diffusion process. By observing with a polarizing
microscope the scattering volume of the smectic A sample after
exposure to the exciting fringe grating some distorted regions
appear with a pattern that remembers the applied heating profile
(fig. 4). These regions are characterized by permanent localized
defects as dislocations or focal conics: this means that before
creating defects inside the layered structure we have also
excited buckling instabilities. If the applied stresses, connected
with the transverse thermal gradient, is not too high the deformed

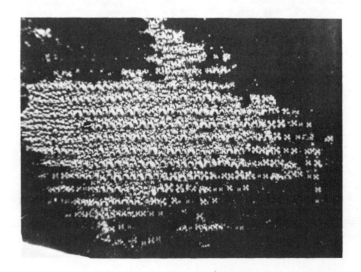

Fig. 4.   Defects in S_A sample: the excitation fringe pattern is
          evident.

structure relaxes elastically but if the deformation is high some localized defects can be also created. In such conditions we observe that the scattered light intensity connected with the elastic relaxation process can increase, due to the presence of the defects, and it masks completely the thermal diffusion effect. As a consequence of the previous preliminary observations we believe that the slow decay time τ_2 might be attributed to the refractive index changes connected with elastically relaxating deformations.

The peculiar elastooptical properties of smectic A materials could be analysed with a forced Rayleigh scattering technique because in this way a defects pattern with controlled periodicity is obtained.

References

1. N. A. Clark and R. B. Meyer, Appl.Phys.Lett, 22:493 (1973).
2. R. Ribotta, G. Durand and J. D. Litster, Solid State Comm., 12:27 (1973).
3. P. G. de Gennes, Solid State Comm., 10:753 (1972).
4. F. J. Kahn, Appl.Phys.Lett., 22:111 (1973).
5. R. Ribotta, Thesis, Orsay 1975.
6. F. Scudieri, T. Papa, D. Sette, M. Bertolotti, E. Sturla, J.Phys., 40 C3:392 (1979).

FORCED RAYLEIGH SCATTERING IN A CRITICAL BINARY LIQUID MIXTURE

Dieter W. Pohl

IBM Zurich Research Laboratory

8803 Rüschlikon, Switzerland

Extremely large and slow signals are found in a forced Rayleigh-scattering (FRS) experiment on a critical mixture of cyclohexane and methanol. They are interpreted in terms of a concentration grating which builds up under the influence of the FRS thermal grating because of the Soret effect. Coupling and decay time tend to diverge at the critical temperature.

Forced Rayleigh scattering (FRS) (1-3) was recently employed by Thyagarajan and Lallemand to create and detect concentration (c) gratings in a binary liquid mixture (4). The c-grating is a manifestation of the Soret effect which was discovered exactly one hundred years ago (5). It accounts for a modification of Fick's law in the presence of a temperature gradient ∇T:

$$\dot{c} = D_m \{ \nabla^2 c - (k_T/T)\nabla^2 T \} \tag{1}$$

k_T being the thermal diffusion ratio and D_m the mass diffusivity. The Soret effect is a direct consequence of the minimum principle for the chemical potential based on the second law of thermodynamics.

The FRS stationary thermal grating ampliture $\hat{T}$ depends on the intensities I_a, I_b of the pump beams (a, b), the absorption coefficient α, the grating constant q, and the heat conductivity K (1,2).

$$\hat{T} = \alpha(I_a \cdot I_b)^{1/2}/q^2 K \tag{2}$$

The stationary amplitude $\hat{c}$ is obtained from Eq.(1) by putting $\dot{c}$ = 0.

$$\hat{c} = (k_T/T)\ \hat{T} \tag{3}$$

The thermal diffusion ratio diverges upon the approach of a critical consolute point – a prediction (6) that was recently confirmed by Giglio and Vendramini (7) using a classical Soret cell and an optical beam deflection technique. The critical temperature T_c was approached to 0.5 K in that experiment.

It is interesting to do a FRS experiment in a binary liquid mixture near its critical point because of the great sensitivity of this technique which allows one to go considerably closer to T_c. Here, I wish to present a few preliminary results of such an investigation using a critical mixture of cyclohexane and methanol (28.6% vol.) The mixture was slightly colored by the addition of some iodine which provided α = 0.093 cm^{-1} at 488 nm, the wavelength of the argon pump laser.

The experimental setup is sketched in Fig.1. It differs from the basic arrangement (1-3) by the use of a separate reference beam similar to Ref. 4. The relative phase between reference (r) and probe (p) beams is adjustable by means of the piezomount PM1. The pump beams (a, b) are not chopped as usual, but undergo a periodic relative phase reversal due to an oscillatory motion of PM2. As a result, the thermal grating also reverses phase for each half period. The concentration grating sluggishly follows this motion with a time dependence.

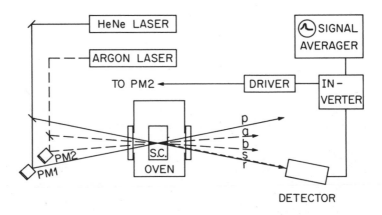

Fig. 1.   Experimental setup, S.C. = sample cell; the other symbols are explained in the text.

$$\hat{c} = {}^{+}_{-}\hat{c}_0 \{1 - 2 \exp(-t/\tau_m)\} \tag{4}$$

The response time

$$\tau_m = (D_m q^2)^{-1} \tag{5}$$

can be adjusted conveniently by the angular separation of the pairs of laser beams which determine q.

The received heterodyne signal (s) has the same temporal behavior as $\hat{c}(t)$ except for a fast contribution from $\hat{T}(t)$. Before recording on a signal averager, the signal gets inverted electronically during each "negative" interval of the thermal grating (8). Non-phase-sensitive perturbations of the signal are eliminated to a large extent in this way. A further advantage of the phase inversion comes from the stationary overall heating by the pump beams. The latter, by the same mechanism which causes the concentration grating, produce a strong thermal lens effect (9). In the present mode of operation, lensing is temporarily constant and cannot disturb the record of the dynamical response.

The experimental setup turned out to be extremely sensitive allowing for pump powers as small as 2 x 1.5 mW at a spot size inside the sample of w = 0.43 mm. Under these conditions T is of the order 10 μK only.

Three typical signals, obtained for $T - T_c \simeq 0.16$, 0.14 and $0.12°C$ and $q = 2812$ cm^{-1} ($\theta_{HeNe} = 28.3$ mrad) are shown in Fig 2. They clearly demonstrate slowing-down and amplitude increase upon approach of T_c (45.30°C in the present setup). The FRS was investigated between 54°C and 45.42°C corresponding to $\varepsilon = T - T_c/T_c = 0.03$ and 0.0005. The decay time varied by a factor of 15 in this interval (fig. 3). It can be fitted to a power law with exponent 0.57 which compares well with the theoretical value for ν, the relevant critical exponent (10), and previous experimental work (11) on lutidine-water mixtures based on classical Rayleigh scattering. Multiple scattering, a serious limitation to classical light scattering near a critical point, did not disturb the FRS signal because of its coherence and directionality.

The increase in signal amplitude being based on absolute intensity measurements, and sensitive to small variations in alignment, scatters considerably more than the decay-time results. They do not yet allow for a quantitative description. $\hat{c}$ can be estimated to be of the order 10^{-6} yielding a k_T between 1 and 10. Comparing the results for three different angles (20.0 x 1, $\sqrt{2}$, $\sqrt{3}$ mrad), strong deviations from Eq. (4) were found but have not yet been analyzed in detail. It appears as if τ_m vs. q decreases slower than q^{-2}.

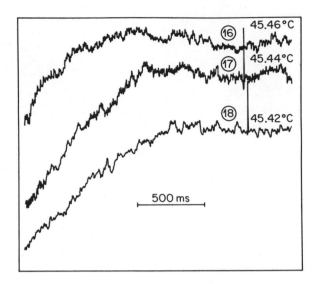

Fig. 2.   Three typical signals close to T_c.

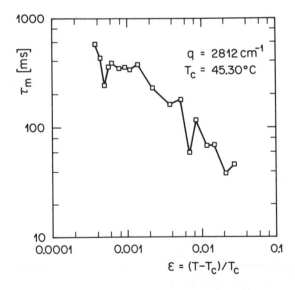

Fig. 3.   Temperature dependence of the response time.

In conclusion our preliminary results are in full agreement with the expectations and indicate a new and convenient way of investigating critical phenomena. We are continuing this work with emphasis on the behavior very close to T_c.

References

1. D. W. Pohl, S. E. Schwarz, and V. Irniger, Phys. Rev. Lett., 31:32 (1973); H. Eichler, G. Salje, and H. Stahl, J. Appl. Phys. (44:5383 (1973).
2. D. W. Pohl, IBM J. Res. Develop. (Review to be published Sept. 1979).
3. F. Rondelez, The capabilities of the forced Rayleigh scattering technique (present proceedings).
4. K. Thyagarajan and P. Lallemand, Optics Comm., 26:54 (1978).
5. C. Soret, Arch. Sci. phys. nat., 2:48 (1879).
6. See, for instance, J. Swift, Phys. Rev. 173:257 (1968); M. Sh. Giterman and E. E. Gorodetskii, Zh. Eksp. Teor. Fiz., 57:637 (1969) (Sov. Phys. – JETP 30:348 (1970)).
7. M. Giglio and A. Vendramini, Phys. Rev. Lett. 34:561 (1975).
8. W. Chan and P. S. Pershan, Phys. Rev. Lett., 39:1368 (1977). Biophys. J., 23:427 (1978).
9. M. Giglio and A. Vendramini, Appl. Phys. Lett., 25:555 (1975).
10. L. P. Kadanoff and J. Swift, Phys. Rev., 166:89 (1968); Ann. of Phys., 61:1 (1970).
11. See, for instance, E. Gülari, A. F. Collings, R. L. Schmidt and C. J. Pings, J. Chem. Phys. 56:6169 (1972).

INDEX